Disclaimer

The publisher of this book is by no way associated with the National Institute of Standards and Technology (NIST). The NIST did not publish this book. It was published by 50 page publications under the public domain license.

50 Page Publications.

Book Title: Smoke Alarm Performance in Kitchen Fires and Nuisance Alarm Scenarios

Book Author: Thomas G. Cleary; Artur A. Chernovsky;

Book Abstract: Tests were conducted to assess the performance of various residential smoke alarms to kitchen fires and nuisance alarm cooking scenarios. A test structure representing a kitchen, living room and hallway was constructed to conduct the tests. Eight different residential smoke alarms types, two photoelectric models, two ionization models, two dual sensor models, and two multi-sensor, intelligent models were used in this study. The data gathered provided insight into the susceptibility of alarm activation from exposures to typical cooking events and alarm times for actual kitchen fires. The effects on the type of alarm, and its distance from the cooking activity or fire were examined. Combustible materials typically found on a counter top can spread flames to overhead cabinets, and a single empty 0.6 m wide 1.0 m tall cabinet can produce a peak heat release rate nearly sufficient to flashover a small room. A protective metal barrier on the bottom and side facing the range tended to limit the spread of flames to the cabinet and reduce the heat release rate. All smoke alarms responded before hazardous conditions developed. The I1 alarm tended to respond first at a given location. Results show smoke alarms placed at the furthest location may provide less than 120 s of available safe egress time, which suggests a more central alarm location closer to the kitchen for this configuration. Ten cooking activities were examined to determine an alarm‰os propensity to activate to cooking aerosols. In most cases, the propensity to nuisance alarm decreased as the distance from the cooking source increased. Alarms that rely on sensitive ionization chambers (here I1 and D2) experience more nuisance alarm activations across all cooking activities and locations. All alarms except I1 and D2 experienced about the same nuisance alarm frequency across all cooking activities for locations outside the kitchen.

Citation: NIST TN - 1784

Keywords: smoke alarms; kitchen fires; detection; nuisance alarms

NIST Technical Note 1784

Smoke Alarm Performance in Kitchen Fires and Nuisance Alarm Scenarios

Thomas G. Cleary
Artur Chernovsky

http://dx.doi.org/10.6028/NIST.TN.1784

National Institute of
Standards and Technology
U.S. Department of Commerce

NIST Technical Note 1784

Smoke Alarm Performance in Kitchen Fires and Nuisance Alarm Scenarios

Thomas G. Cleary
Artur Chernovsky
Engineering Laboratory

http://dx.doi.org/10.6028/NIST.TN.1784

January, 2013

U.S. Department of Commerce
Rebecca M. Blank, Acting Secretary

National Institute of Standards and Technology
Patrick D. Gallagher, Under Secretary of Commerce for Standards and Technology and Director

National Institute of Standards and Technology Technical Note 1784
Natl. Inst. Stand. Technol. Tech. Note 1784, 80 pages (January 2013)
http://dx.doi.org/10.6028/NIST.TN.1784
CODEN: NTNOEF

Abstract

Experiments were conducted to assess the performance of various residential smoke alarms to kitchen fires and nuisance alarm cooking scenarios. A structure representing a kitchen, living room and hallway was constructed to conduct the experiments. Eight different residential smoke alarms types, two photoelectric models (P1 and P2), two ionization models (I1 and I2), two dual sensor photoelectric/ionization models(D1 and D2), and two multi-sensor, intelligent models (M1 and M2) were used in this study. The data gathered provided insight into the susceptibility of alarm activation from exposures to typical cooking events and alarm times for actual kitchen fires. The effects of alarm technology and installation location on the propensity of an alarm to activate were examined. In the kitchen fire experiments, all smoke alarms responded before hazardous conditions developed. An ionization alarm (I1) tended to respond first compared to other co-located alarms. Results show smoke alarms placed greater than 6 m from the kitchen range may provide less than 120 s of available safe egress time, which suggests the importance of a more central alarm location closer to the kitchen for this configuration. Experiments were conducted to determine an alarm's propensity to activate when exposed to particulates generated from eight typical cooking activities including toasting, frying, baking and broiling. In most cases, the propensity to nuisance alarm decreased as the distance from the cooking source increased. Two alarms, I1 and D2, experienced more nuisance alarm activations across the eight cooking activities than the other alarms. The remaining alarms experienced about the same combined nuisance alarm frequency by averaging all cooking events for installation locations outside the kitchen. Experiments showed combustible materials typically found on a counter top can spread flames to overhead cabinets, and a single empty 0.6 m wide 1.0 m tall wood-framed, pressboard cabinet can produce a peak heat release rate nearly sufficient to flashover a small room. Alternatively, protective metal barrier on the bottom and side facing the range tended to limit the spread of flames to the cabinet and reduce the heat release rate.

Disclaimer

Acknowledgements

The authors acknowledge Michael Selepak, Anthony Chakalis, Doris Rinehart, Jay McElroy, and Laurean Delauter for their assistance with the set up and gathering data. We also want to thank the Montgomery County Fire and Rescue Service for the use of the burn prop building at the Public Safety Training Academy, and Assistant Chief Mike Clemens and Captain Audrey Deputy for their assistance.

This project was funded by an interagency agreement with the US Consumer Products Safety Commission. The technical monitor of the contract was Mr. Arthur Lee.

Table of Contents

List of Figures

List of Tables

1   Introduction

According to the National Fire Protection Association (NFPA), there were 1 451 500 fires reported in the United States during 2008 [1]. These fires caused 3320 civilian fire deaths, 16 705 civilian fire injuries, and $15.5 billion in property damage. Homes with working smoke alarms typically had a fire death rate that is about half the rate for homes with no smoke alarms or with alarms that failed to operate [2]. A 2008 telephone survey found that 96 % of U.S. households reported having at least one smoke alarm [3]. Despite this reported high coverage, between 2003 and 2006 two out of five homes (41 %) in the reported home fires had no smoke alarms or had smoke alarms that failed to operate properly. Telephone polling of US households conducted in 2010 for the NFPA reported 52 % of all respondents that had at least one smoke alarm indicated there was a smoke alarm installed in the kitchen; 43 % of the households reported a nuisance alarm within the last year; and, of that 43 % about 75 % reported they thought nuisance alarms were caused by cooking activities [3]. Various studies have previously examined the likelihood of smoke alarms to remain operational after installation and have identified the type of smoke alarm (photoelectric or ionization) and the location of smoke alarms to cooking sources [4-6].

Cooking appliances are the leading ignition sources in home fires and cause an average of 150 000 home structure fires a year, leading to 500 deaths and 4660 injures (2003 to 2006 yearly average) [7]. Fires caused by un-attended cooking or un-supervised children cooking can grow rapidly; thus, early detection from working smoke alarms is critical. Further complicating matters, since smoke alarms are prone to nuisance alarms from cooking particulates they are subject to intentional power disconnection or removal by consumers. These statistics raise questions for both consumers and smoke alarm experts: What type of smoke alarm should be installed and how far away from cooking appliances (the origin of the nuisance aerosols) should an alarm be installed to reduce the frequency of nuisance alarms while still providing a high detection capability for kitchen fires? To answer these questions, specific information is needed, such as: How fast do kitchen fires grow? How quickly do hazards develop? What are the characteristics (primarily particle size distribution and concentration) of nuisance source aerosols for typical cooking activities? What are the performance characteristics of smoke alarm technologies to kitchen fires and nuisance aerosol exposures from cooking activities in terms of adequate detection of fire and acceptably low nuisance alarm activations? Are there any new detection technologies that will improve the situation, and how will new technologies be evaluated?

There is little information on smoke alarm activation as it relates to fire growth and hazard development of kitchen fires. A previous NIST smoke alarm study conducted kitchen fire experiments where a pot of cooking oil was heated on a gas range until it ignited [8]. In the four tests conducted the cooking oil ignited after between (20 and 30) min of heating. Both photoelectric and ionization smoke alarms activated at least 10 min prior to ignition of the cooking oil. It was also found that heated oil tended to fill the house with oil particle aerosols well before igniting. In another study Mealy *et al.* conducted two kitchen cabinet fire tests as part of a National Institute of Justice grant [9]. They observed a minimum available safe egress time of greater than 135 s for each alarm evaluated.

NIST conducted nuisance alarm tests as part of the Home Smoke Alarm study [8]. It was observed that nuisance alarms in residential settings from typical cooking activities, smoking, or candle flames are affected by the properties of the aerosol produced and its concentration, the location of an alarm relative to the source, and the air flow that transports smoke to an alarm. The study provides a detailed set of data that addresses several issues involving nuisance alarms and that reinforces current suggested alarm placement practices. The study confirms the practice that alarms not be installed close to cooking

1

appliances if at all possible. The results suggested that nuisance alarms could be substantially reduced by moving the location of an alarm that frequently experiences nuisance alarms well away from cooking appliances while at the same time keeping the alarm within the area to be protected. It was also observed that ionization alarms had a propensity to alarm when exposed to nuisance aerosols produced in the early stages of some cooking activities, prior to noticeable smoke production. This phenomenon could be particularly vexing to homeowners who experience such nuisance alarms.

The Consumer Product Safety Commission, CPSC, conducted an experimental study of the frequency and causes of residential cooking nuisance alarms by monitoring several smoke alarms near kitchens in 9 households for 30 days [10]. Photoelectric, ionization, and dual sensor photoelectric/ionization alarms with disabled sounders were monitored and alarm times were recorded. Additionally occupants were instructed to keep a record of cooking activities and any time the existing household alarms activated. The results showed a considerable reduction in nuisance alarms as the distance from the cooking appliance increased from 1.5 m to 6.0 m. Dual-sensor alarms tend to alarm more frequently than photoelectric or ionization alarms. Additionally, certain types of cooking activities like sautéing, pan frying, and stir frying tended to cause more nuisance alarms than other types of cooking.

The National Fire Alarm and Signaling Code, NFPA 72, addresses the issue of nuisance alarms in household smoke alarms by specifying alarm location rules within 6 m (20 ft.) of horizontal distance as measured from a ceiling location above a fixed cooking appliance to the smoke alarm [11]. Simply stated, according to NFPA 72, no smoke alarms should be located within 3 m (10 ft.) of a ceiling location above a fixed cooking appliance, and between 3 m and 6 m, smoke alarms must use photoelectric detection, or have a means of temporarily silencing the alarm. An exception is specified for placement of photoelectric smoke alarms within 1.8 m (6 ft.) where the 3 m exclusion would prohibit placement of a smoke alarm required by other sections of the code. These rules were made based upon a limited amount of research with the overarching premise that some decrease in nuisance alarms and subsequent decrease in alarm disabling would ultimately improve safety. The efficacy of these location rules on the balance between reduction of nuisance alarms and adequate detection of kitchen fires needs to be further studied. Quantitative evaluation of smoke alarm performance in relation to cooking nuisance source rejection, would verify expected improvement when the location rules are followed. In addition, advances in smoke alarm technology have led to new products that have been designed to mitigate detection and nuisance alarm problems. These products have no measured performance history regarding nuisance alarm rejection. The performance of any new product designed to perform within the 6 m (20 ft.) needs to be studied.

The research presented in this report focuses on alarm performance after exposure to various cooking nuisance sources and cooking fire scenarios. Existing alarm technologies and newer advanced smoke alarms were included in the research. To provide some insight into how fast certain kitchen fires grow, and how quickly hazards develop, the hazard development of kitchen fires that start slowly and grow to involve an overhead cabinet was studied.

2   Experimental Plan

The experimental plan consisted of measuring the sensitivity of various residential smoke alarms inside the NIST fire emulator/detector evaluator (FE/DE). Each smoke alarm was subjected to nuisance source exposures from cooking activities and their propensity to nuisance alarm was documented. Each smoke alarm was also subjected to kitchen fire scenarios to assess their performance in kitchen fires. Nuisance source exposures and kitchen fire tests were conducted in a small apartment mock-up. While this mock-up living space simulates only a fraction of households, it provided significant data regarding nuisance alarms and hazard development during fires due to its relatively small square footage. Additionally, tests were performed in the furniture calorimeter in NIST's National Fire Research Laboratory to measure the heat release rate of two ignition scenarios and various cabinet constructions. These tests were conducted to characterize the burning behavior of the kitchen fire scenarios in over-ventilated conditions. While the limited space of the small apartment mock-up will affect the burning behavior after some period of time due to limited oxygen, these experiments bound expected heat release rate.

The smoke alarms that were tested were selected from current retail stock. A variety of smoke detectors were used this experiment including those employing photoelectric or ionization single-sensor technology, and photoelectric and ionization dual-sensor technology, and multi-sensor, intelligent alarm technology1. Table 1 lists the technology for each alarm and the identifying notation used throughout the rest of the report. Two separate models of each alarm technology were included in this study, and are distinguished by the numeral 1 or 2 in the notation.

Technology	Notation
Photoelectric	P1
Photoelectric	P2
Ionization	I1
Ionization	I2
Dual sensor photoelectric/ionization	D1
Dual sensor photoelectric/ionization	D2
Multi-sensor, intelligent alarm	M1
Multi-sensor, intelligent alarm	M2

Table 1. Alarm technology and identifying notation used in this report.

2.1   Smoke Alarm Sensitivity Test Protocol

The NIST fire emulator/detector evaluator, FE/DE, was used to expose smoke alarms to smoldering cotton wick smoke at various concentration levels. A schematic of the FE/DE is shown in Figure 1. The cotton wick is the same material used in UL 217 for smoke alarm sensitivity test [12]. The FE/DE cotton wick igniter was used to provide stepwise concentrations of smoke. At the test section a laser light extinction beam (635 nm wavelength) located 5 cm below the duct ceiling, traveling the width of the duct, and reflected off mirrors to increase the path length through the duct smoke, was used to measure the light extinction of the smoke. A reference measuring ionization chamber (MIC) was installed on the ceiling of the test section. The MIC responds to smoke in a manner similar to ionization

---

1 Intelligent alarm technology is distinguished by the use of an algorithm to process sensor signals to determine the alarm condition. The intelligent alarms currently available pair ionization sensors with carbon monoxide gas or humidity sensors.

chambers inside smoke alarms. The MIC output current is reduced when smoke is present, and the reduction is related to the smoke concentration. The output current is nominally 100 pA in clean air.

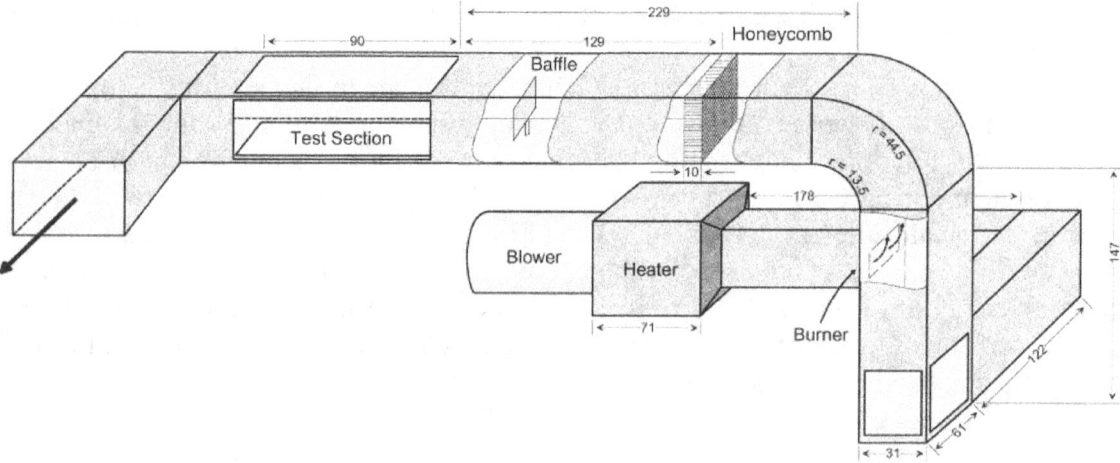

Figure 1. Schematic of the NIST fire emulator / detector evaluator (units in cm).

Since the FE/DE is a single-pass flow device, the smoldering smoke does not get a chance to age for a long time before it reaches the test section. In contrast, the UL217 smoke box has a re-circulating flow path, thus the smoke is aged to some degree. Aging affects the average smoke particle size which in turn affects alarm response. The two dashed curves presented in Figure 2 show the smoke sensitivity test limits specified by the UL 217 Standard in terms of a comparison of the light transmittance through a 1.5 m path length of smoke to the MIC output current. The area between the curves represents expected smoke characteristics. That is, during a sensitivity test, all measured values of light transmittance and MIC current must fall within the bounds of the two curves to have a valid test. Typical measures for the FE/DE cotton wick smoke are shown in Figure 2 as averaged steady smoke values with error bars representing ± one standard deviation for both transmittance and MIC current averages. The averaged values fall within the valid region until the MIC value reaches about 50 pA.

The smoldering cotton produces carbon monoxide (CO) in addition to smoke particles. Figure 3 shows the CO concentration as a function of the MIC current for different steady wick burning periods as measured at the FE/DE test section. A non-dispersive infrared carbon monoxide gas analyzer was used to measure the CO concentration from gas samples extracted from the FE/DE test section through a sampling line. The analyzer has a resolution of 1×10^{-6} volume fraction (ppm volume), and an uncertainty on that order. The plotted error bars represent ± one standard deviation of the fluctuating measurements. For intelligent multi-sensor smoke alarms that use CO sensing, the CO concentration in the smoke sensitivity test may impact alarm conditions and thereby the smoke concentration at alarm.

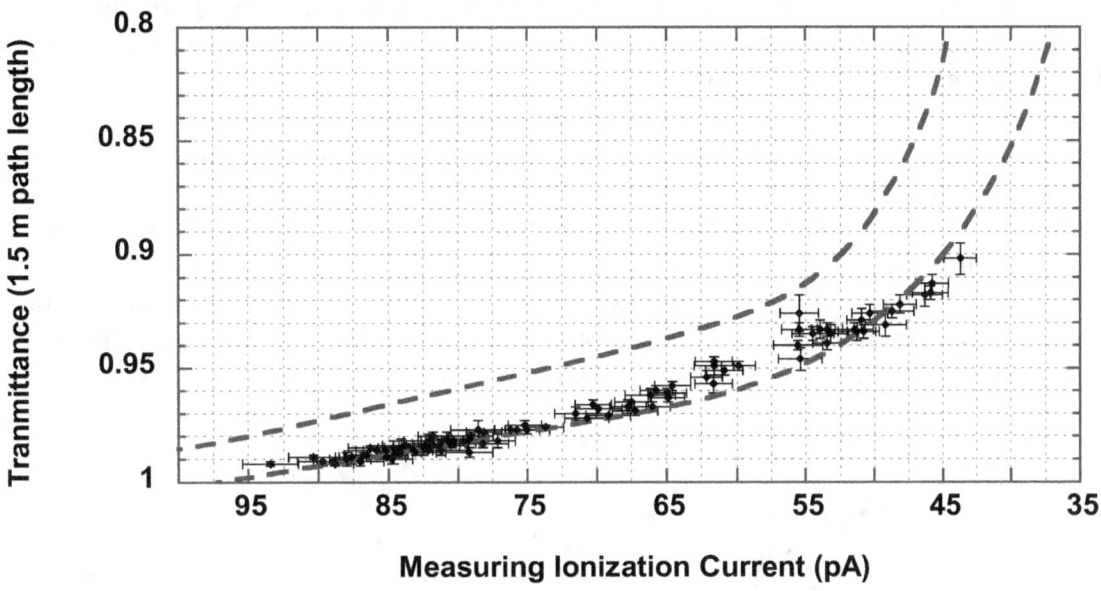

Figure 2. Smoke sensitivity test limits for UL217. Data points are measured values from the FE/DE smoldering cotton smoke (uncertainty bars are 1 standard deviation).

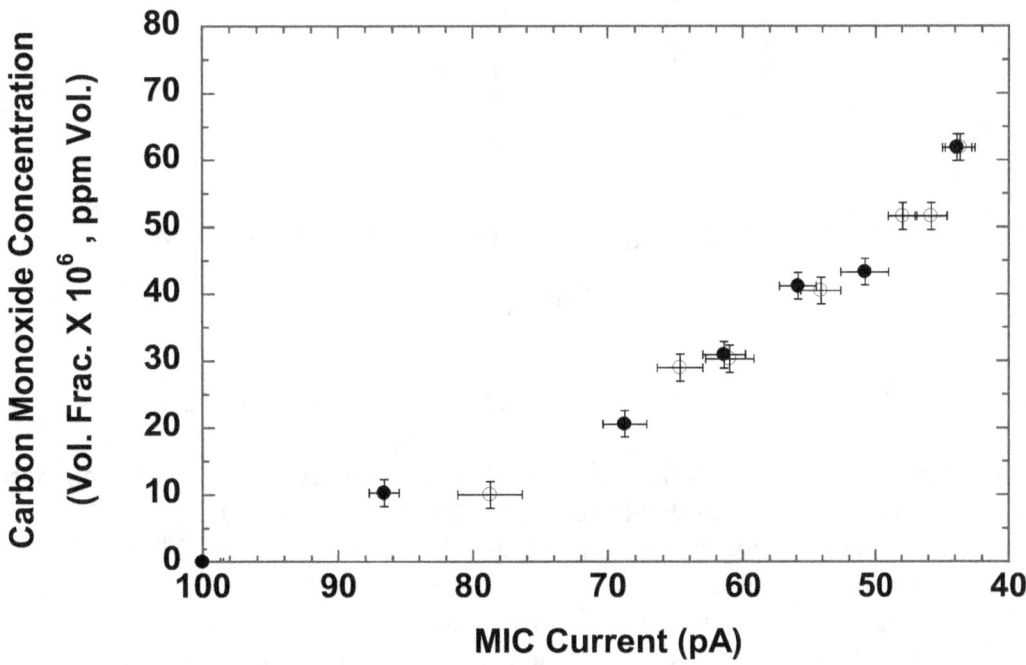

Figure 3. CO concentration versus MIC current for steady cotton wick burning periods. Open and closed symbols represent different sets of wicks (uncertainty bars are 1 standard deviation).

Identical make and model smoke alarms were placed side-by side on the ceiling of the test section, just behind the extinction beams, and 15 cm in front of the MIC. The sensing chambers of the installed smoke alarms were orientated between the best and worst case orientation for smoke entry. All smoke alarms were powered by battery, and the smoke alarm battery voltage was used to determine if an alarm was activated. The alarm state was determined by the smoke alarm battery voltage drop and compared to the smoke extinction and measuring ionization chamber results.

The mid-point between the non-alarm and alarm smoke extinction or measuring ionization chamber values is used as the estimate of the alarm sensitivity. For example, Figure 4 shows a typical graph of MIC current and the laser beam transmittance versus time for an ignition sequence of 6 sets of wicks. During ignition of a set of wicks, the smoke production is elevated and the MIC current and transmittance drop sharply. The wicks in the ignited set then approach a steady burning rate and both the transmittance and the MIC current reach a plateau. Consecutive sets of wicks are ignited and add to the smoke concentration as the previously ignited wicks continue to burn.

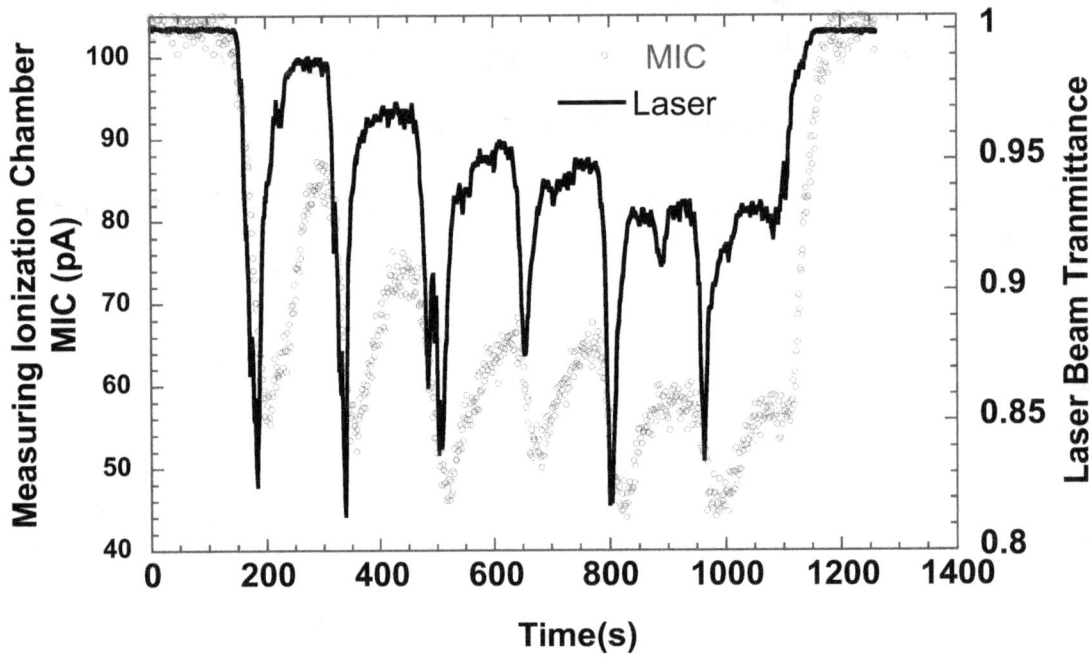

Figure 4. MIC current and laser transmittance for an ignition sequence of six sets of wicks.

Figure 5 shows an expanded view of the time period illustrated in Figure 4. Assuming an alarm was not active prior to the ignition of the 5th set of wicks and was active prior to the ignition of the 6th set of wicks, a midpoint value of the MIC current or transmittance between the 4th and 5th set of wicks just prior to ignition of the next set of wicks is used to estimate the alarm sensitivity. Interval 1 (30 s prior to ignition of the 5th set of wicks) has an average MIC current of 64.4 pA with a standard deviation of 1.2 pA, and an average transmittance of 0.948 with a standard deviation of 0.001. Interval 2 (30 s prior to ignition of the 6th set of wicks) has an average MIC current of 57.6 pA with a standard deviation of 1.5 pA, and an average transmittance of 0.930 with a standard deviation of 0.003. Thus, the average MIC current is 61.0 pA, and the average transmittance is 0.939. Repeated sensitivity test results are averaged.

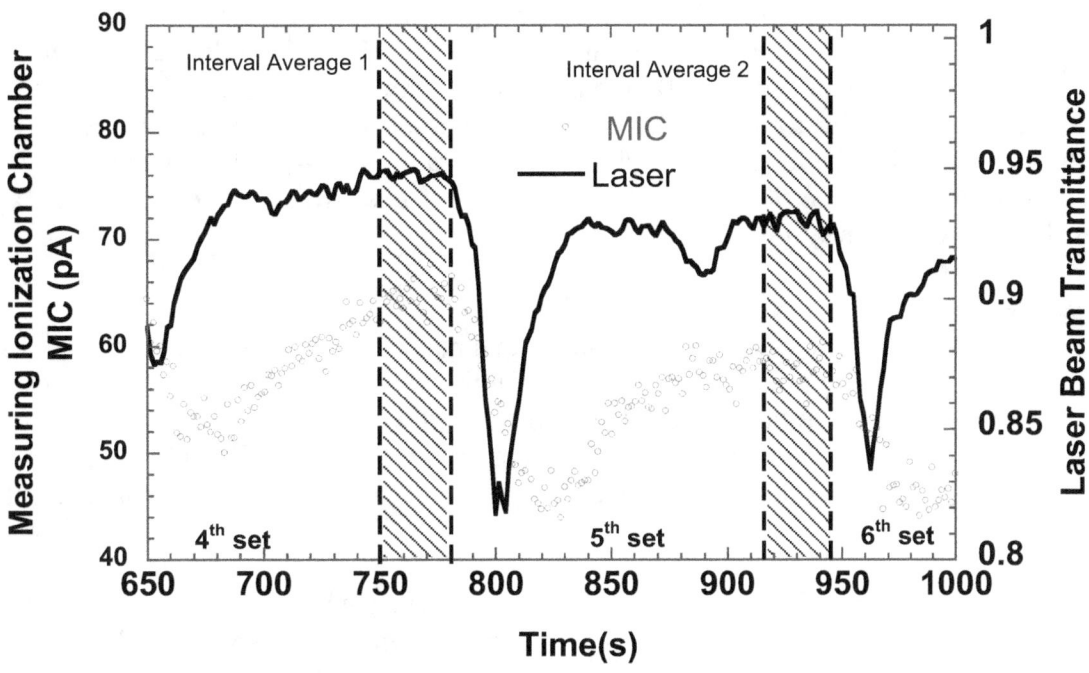

Figure 5. An expanded view of the MIC current and laser beam transmittance for the 4th and 5th set of wicks. The interval averages 1 and 2 represent the steady wick smoke concentration levels.

2.2 Fire Scenario Designs and Heat Release Rate Measurement

Range top initiated fires are the most prevalent residential kitchen fires according to the US national fire loss statistics [7]. Un-attended range top cooking fires can initiate and grow unnoticed prior to a smoke alarm alert or discovery by an occupant. Food items are the most-likely materials first ignited in a range top fire, followed by fire spread to adjacent items such cabinets or combustible items on counters. Extended heating of solid food to the point where it chars and ignites, or heating cooking oil until it reaches its fire point, are fire initiation events. These types of fire initiations, however, may not represent significant challenges to smoke alarms because they tend to have an extended production of smoke prior to ignition. That smoke tends to activate local smoke alarms well before hazardous conditions develop. A more challenging fire scenario is direct ignition of combustibles from a stove-top heating element, since smoke production and the fire essentially begin at the same time. The fire scenario used in this experiment started by the ignition of a roll of paper towels on the counter adjacent to the range heating element, followed by subsequent ignition of various items on the counter top (such as plastic plates, boxed cereal, plastic coffee brewer, etc.) and fire spread to an overhead wall cabinet.

Two cabinets and two ignition scenarios were investigated. The two cabinets were identical in size, 61 cm wide by 76 cm high, by 30 cm deep (30 in. x 24 in. x 12 in.) but with different materials of construction. The first cabinet was unfinished and had a solid oak frame with oak door panels and pressboard top, bottom, interior shelf and side panels. The second cabinet was constructed from pressboard with a thin plastic veneer finish. It contained one interior shelf. During all tests, the cabinets were empty except for the shelf board.

7

The ignition scenarios consisted of two different fixed arrangements of combustible materials. The first arrangement consisted of a roll of paper towels sitting on a stack of five 25 cm diameter foamed polystyrene disposable plates, adjacent to a 300 g bag of potato chips, and a small plastic electric drip coffee maker. Figure 6 shows the arrangement of the combustibles underneath the cabinet and adjacent to the range. For the fire tests, the range was replaced with a frame of cement board and a 1kW electric heating element to simulate an electric range. The roll of paper towels was unraveled and the paper towel end was draped over the heating element.

The second arrangement (Figure 7) consisted of a roll of paper towels sitting on a stack of ten 25 cm diameter foamed polystyrene disposable plates, adjacent to a bag of corn chips, a box of breakfast cereal, a bag of potato chips and a box of microwave popcorn. On the counter in front of the paper towels was a rigid plastic plate with five paper towels on top that were soaked with 100 ml of cooking oil. In addition a cotton rag soaked with 50 ml of cooking oil was draped over the counter and on the range mock up. Identical to the first ignition scenario, the roll of paper towels was unraveled and the paper towel end was draped over the heating element.

The ignition sequence was initiated by applying power to the electric heating element. Once the heating element reached a high enough temperature, the paper towel end ignited and spread to the entire roll, flames spread to the different combustibles and eventually impinged on the bottom of the cabinet.

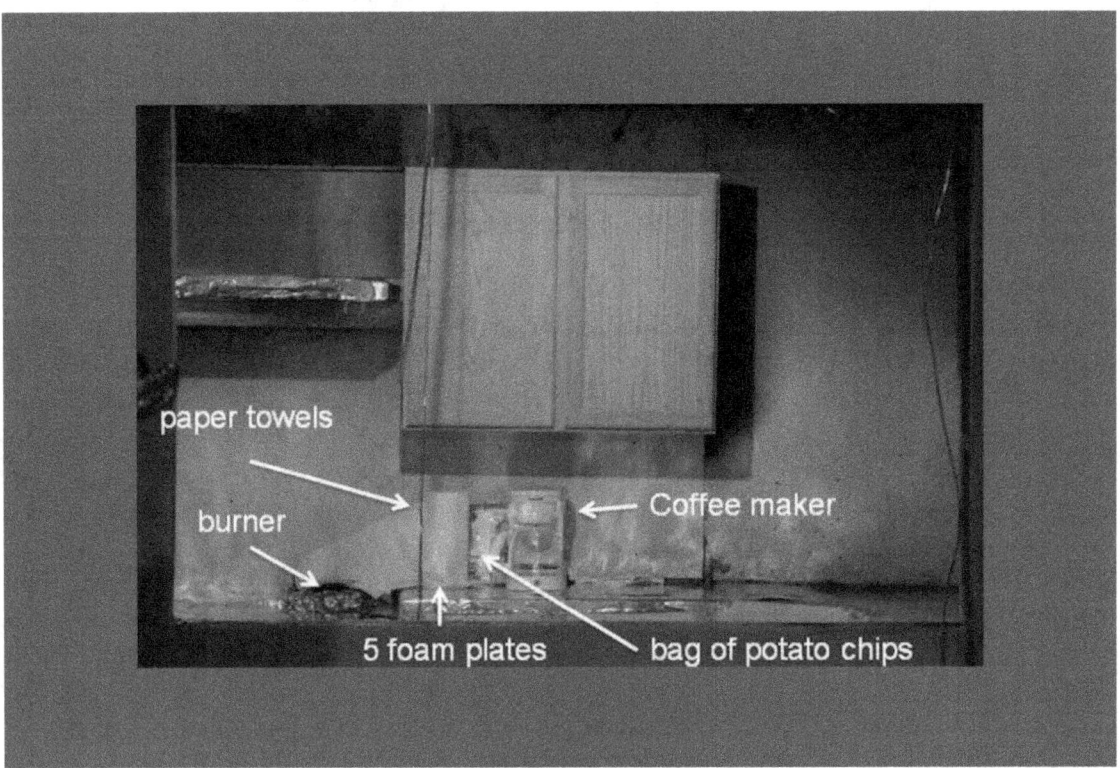

Figure 6. Configuration for ignition scenario 1.

Figure 7. Configuration for ignition scenario 2.

Additional tests were conducted with a sheet metal barrier placed on the bottom and partially up the side of the cabinet facing the range. The intent of the barrier was to protect the cabinet from impinging flames in order to slow down or eliminate the ignition of the cabinet. Figure 8 shows how the sheet metal was installed on the cabinet. This protective layer was intended to simulate an aesthetically pleasing barrier built into the cabinet. The kitchen fire tests used aluminum sheet metal, while the furniture calorimeter tests used a galvanized steel sheet metal barrier.

Figure 8. Sheet metal barrier on bottom and side of wall cabinet.

A portable mockup of the kitchen section was set up under 3x3 m furniture calorimeter hood in the NIST National Fire Research Laboratory. The furniture calorimeter has a 1MW capacity and was calibrated with a natural gas calibration burner prior to each series of tests (4 tests per day) The standard 5 point natural gas calibration is performed at 75/150/200/350/500 kW fuel flow presets to determine calibration factors. The combined standard uncertainty of heat release rate for an unspecified fuel was estimated as ± 8 %, and the combined standard uncertainty of the total heat release was estimated as ± 5 % due to the uncertainty in the heat of combustion of mixed fuel items [13]. Fire resistant cement board panels were used to create the counter top, the supporting back wall, a simulated range cabinet over the range hood, and the ceiling section. Gypsum board was attached to the supporting back wall, and the cabinet was attached to the gypsum board. Figure 9 shows the arrangement. The gypsum board sections were replaced after each test. Tests were also conducted with a cement board mock-up of the cabinet to assess the heat release rate without the cabinet.

Figure 9. Kitchen counter and cabinet mockup. The counter level rests on load cells, and the entire mockup fits under the furniture calorimeter hood.

2.3   Full-scale Tests

2.3.1 Test Structure

Full-scale tests were conducted at the Montgomery County Fire and Rescue Service Public Training Academy. A section of the burn prop building (Figure 10) was used to conduct the experiments.

Figure 10. Exterior view of the burn prop building.

A kitchen, living room, and hallway mock-up was arranged in a section of the first floor of the burn prop building. Figure 11 is a schematic of the mock-up. In this configuration, the hallway is leads to additional rooms. The opening on the wall was an access doorway into the structure. There was another door opening from the kitchen to the outside of the burn prop building that was used to ventilate the mock-up to the outside after each test. The kitchen has two access openings and a wide window-style opening looking out into the living room. All three openings had the same soffit depth from the ceiling (30 cm). The schematic shows the location of thermocouple trees (TC Tree), gas sampling (Gas Analyzer), and Laser Extinction meters (Laser). Figure 12 is a picture of the kitchen layout looking through the kitchen/living room opening.

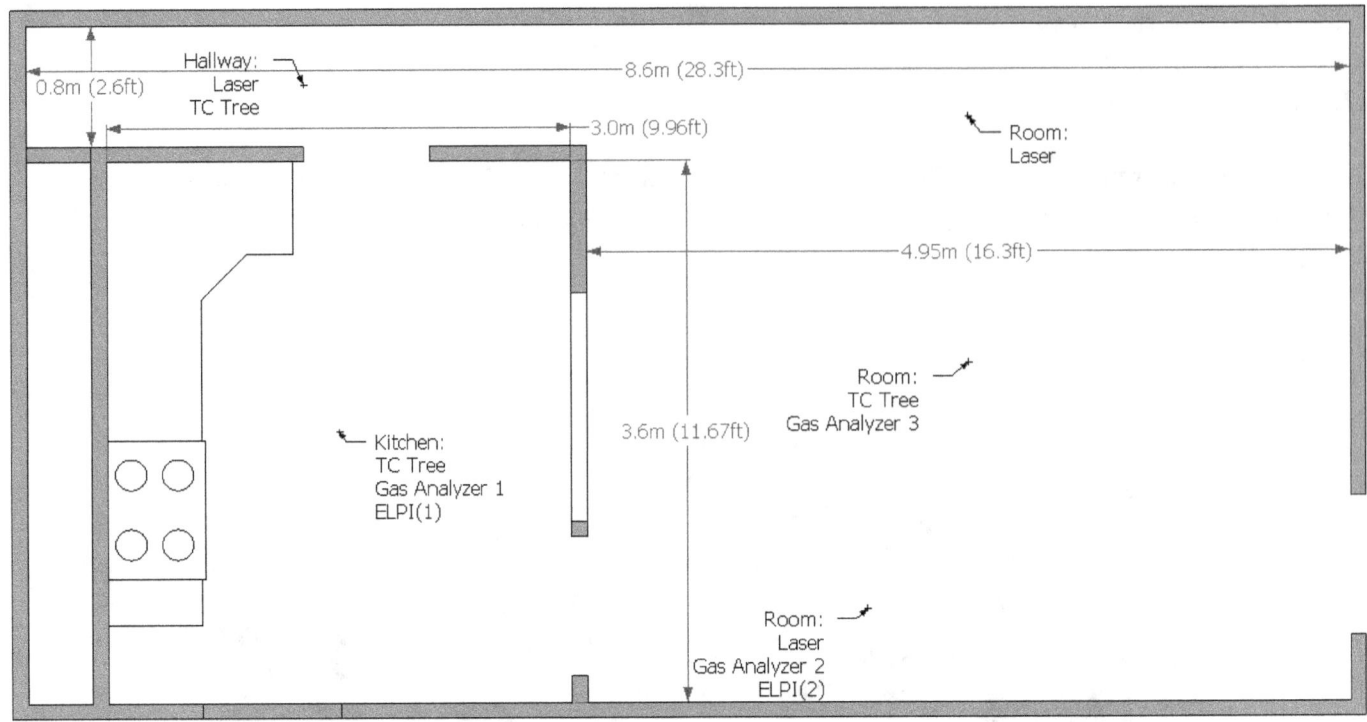

Figure 11. Schematic of the living space mock-up.

Figure 12. Picture of the kitchen counter and cabinet mock-up.

2.3.2 Measurement Equipment

The mock-up was instrumented with gas sampling tubes, thermocouples, laser extinction meters, and smoke alarms that were monitored for alarm state. Figure 13 shows the view looking from the access door into the kitchen. The positioning of three smoke alarm boards is shown. The individual smoke alarms are obscured. A laser extinction meter and a gas sampling tube are visible below the smoke alarms. The laser extinction meter and the gas sampling tube were positioned at 1.5 m from the floor, a standard height for tenability evaluation. The combined standard uncertainty of the laser extinction meter was estimated as $\pm$ 10 % of the recorded optical density. The combined standard uncertainty of both the CO and CO_2 gas concentration measurements was estimated as $\pm 5\times10^{-4}$ volume fraction.

Figure 13. Picture showing alarm placements, extinction meter, sampling tubes, and window and door openings.

The alarm state of each smoke alarm was estimated from battery voltage measurements. Each smoke alarm shows a distinct drop in the battery voltage when the buzzer is sounding. This voltage drop is indicative of a sounding alarm. The estimated uncertainty in the reported time to alarm is $\pm$ 1 s.

13

2.3.3 Nuisance Alarm Test Protocols

Cooking activities such as toasting, frying, baking and broiling were selected to represent a range of potential cooking sources that could trigger nuisance alarms. The CPSC study [10] guided the selection of the sources. These sources were also used in additional cooking source experiments at NIST [14].

Toasting bread

The toasting bread experiment consisted of two slices of white bread placed into a two-slice toaster. The automatic pop-up function of the toaster was disabled. Two slices of white sandwich bread were placed in the toaster and 120 s after the start of the data acquisition computer, power was applied to the toaster. The bread was toasted for a fixed period of time, and then the toaster was powered off. Three separate toasting times were specified 105 s, 185 s and 220 s representing light, dark, and very dark toast (burnt), respectively. No one was in the test room during these experiments. Figure 14 shows the location of the toaster on the counter and the representative toasted bread samples for the three toasting times.

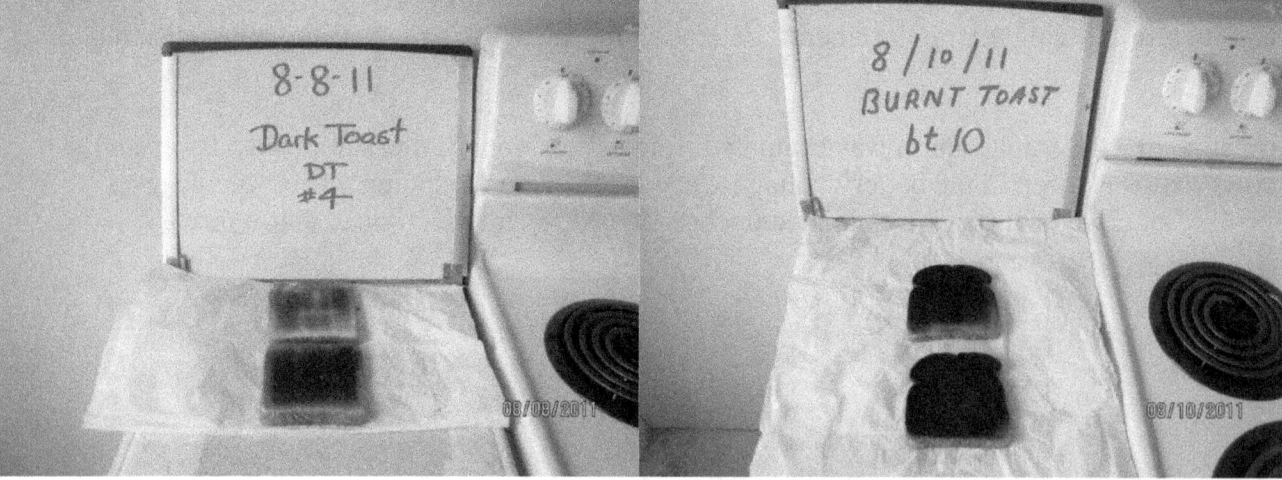

Figure 14. Toasting bread configuration and toasted bread exemplars.
Toasting bagel

The toasted bagel experiments consisted of one regular frozen bagel cut in half. Each half toasted in the two-slice toaster. The automatic pop-up function of the toaster was disabled. The bagel was toasted for 240 s then the toaster was powered off.   Figure 15 shows a representative sample of a toasted bagel.

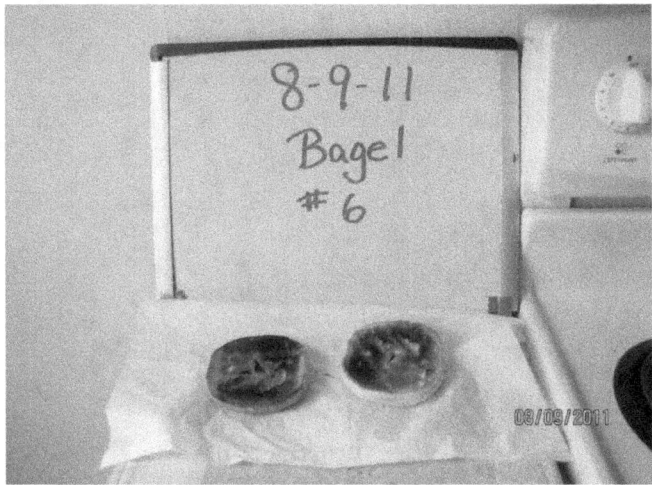

Figure 15. Toasted bread exemplar.

Frying bacon

The frying bacon experiment consisted of frying six strips of bacon in a 25 cm diameter nonstick-coated frying pan on a 19 cm diameter 1.1 kW electric coil burner on the range. The range burner was turned on to the highest heat setting for 60 s after the start of the data acquisition computer.   The bacon was stirred and turned for the next 380 s, fully cooking the bacon to a crispy texture.   The frying pan was removed from the range and the heat turned off. Figure 16 shows representative before and after images of the bacon.

Figure 16. Frying bacon configuration and fried bacon exemplar.
Frying hamburger

The fried hamburger experiment consisted of one frozen beef hamburger patty placed in a 25 cm fry pan and heated on a 19 cm diameter 1.1 kW electric coil burner on the range. The coil burner on the range was set to the high heat setting (10) and the frying pan with the hamburger was placed on the burner. After 180 s the heat was reduced to a medium setting (6) setting, 150 s later, it was flipped. The hamburger was allowed to cook for an additional 180 s, at which time the heat was shut off and the frying pan removed from the range. Figure 17 shows before and after images of the hamburger.

Figure 17. Frying hamburger configuration and fried hamburger exemplar.

Broiling hamburger

The broiling hamburger experiment consisted of broiling a frozen beef hamburger patty using a broiler pan placed on the top oven rack of an electric range. The broiler pan with the hamburger was placed in the oven with the oven door was left cracked approximately 11.5 cm and the oven was set to broil. After 600 s the oven door was opened and the hamburger was flipped. The door was then returned to its cracked open position and the hamburger was left to broil another 240 s. The hamburger and broiler pan were removed and the broiler turned off. Figure 18 shows before and after images of the patty.

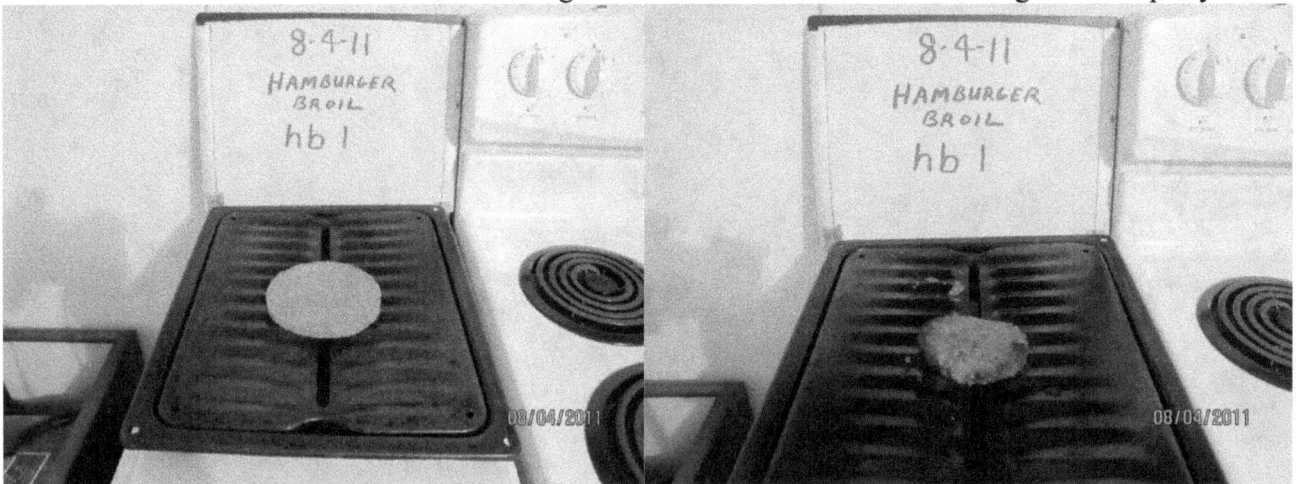

Figure 18. Broiling hamburger configuration and broiled hamburger exemplar.

Grilled Cheese Sandwich

The grilled cheese sandwich experiment consisted of two slices of white sandwich bread, buttered on the outside, with two slices of American cheese inside, placed in a 25 cm diameter frying pan and heated on a 19 cm 1.1 kW electric coil burner on the range. The coil burner on the stove was set to the high heat setting (10) and the frying pan with the sandwich in it was placed on the burner. After 180 s the heat was reduced to a medium high setting (7), and the sandwich was flipped over. The sandwich was allowed to cook for another 100 s at which time the heat was shut off and the frying pan removed from the range. Figure 19 shows the set up and the prepared sandwich.

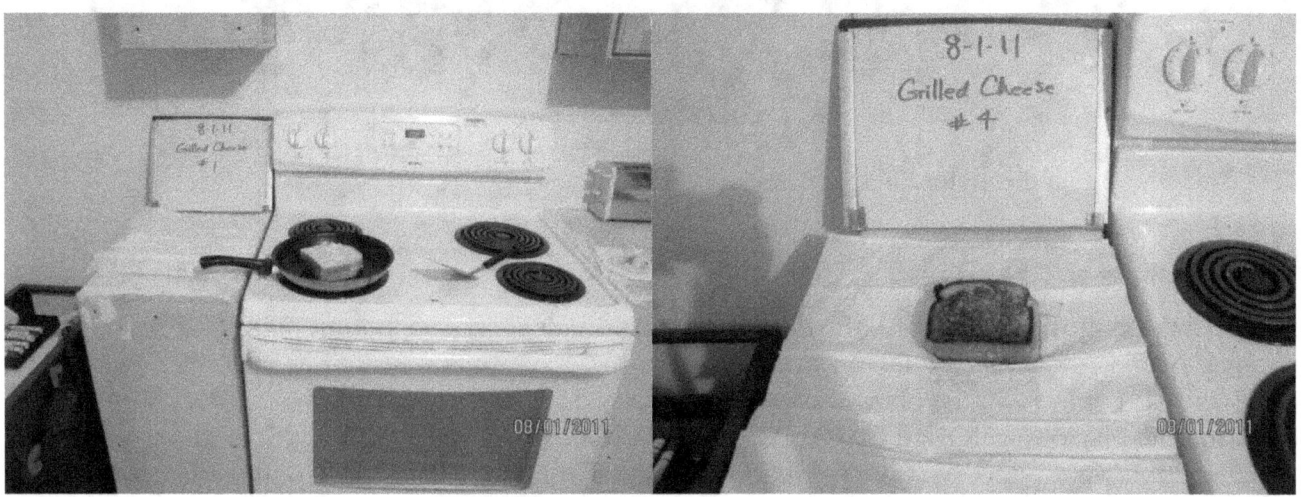

Figure 19. Grilled cheese sandwich configuration and prepared sandwich exemplar.

Vegetable Stir Frying

The vegetable stir frying experiment consisted of chopping-up one carrot, one onion and one celery stalk and frying them in a 27.5 cm diameter steel wok pan with 10 ml of vegetable oil. After 60 s of data collection, 15 ml of vegetable oil was poured into the wok pan on the front 19 cm coil burner which was then set to high heat setting (10). After heating the vegetable oil 140 s the carrots, onions and celery were stirred together in the wok pan. The vegetables were continuously stirred for 165 s at which time the heat was turned down to a medium setting (6). Stir frying continued for 140 s longer then the wok pan was removed from the range. Figure 20 shows the chopped vegetables before and after cooking.

Figure 20. Stir fry vegetables before and after cooking.

Baking Pizza

The baking pizza experiment consisted of baking a small individual size pepperoni pizza (6.5 oz.) in the electric range oven. Prior to placing the pizza in the oven, the oven was preheated to a setting of 450° F. The oven door was opened and the pizza was placed directly on the mid-level oven rack. The oven door was closed and the pizza was allowed to bake for 600 s. At the end of 600 s cooking time, the oven door was opened and the pizza removed. The oven door remained open for a total of 30 s then the door was closed and the oven turned off. Figure 21 shows the pizza before and after cooking.

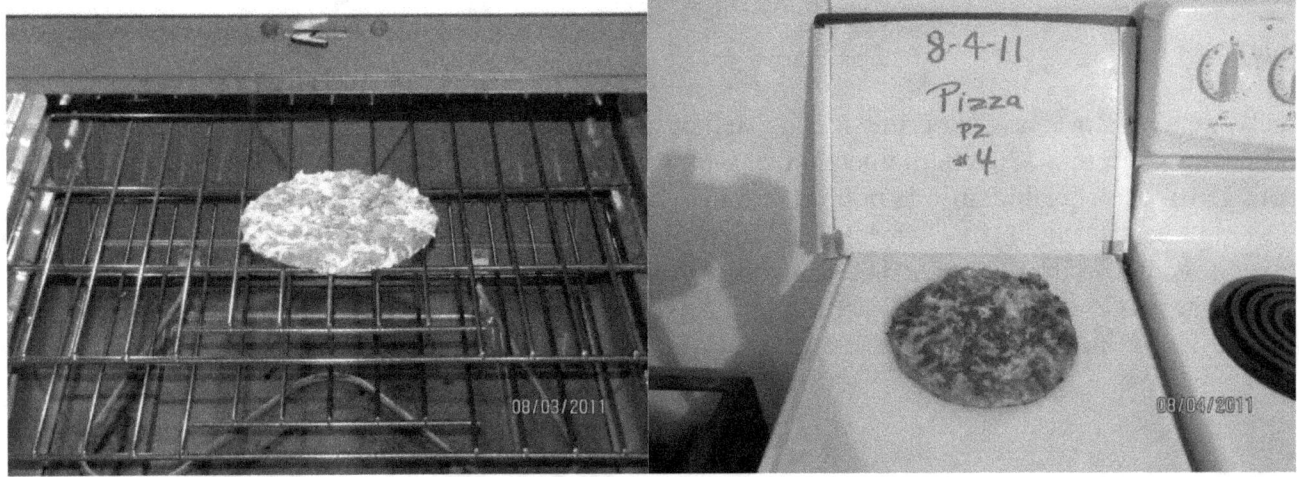

Figure 21. Baking pizza configuration and cooked pizza.

2.3.4 Kitchen Fire Experimental Protocols

The kitchen fire experiments used the same fire scenarios tested in the furniture calorimeter, namely the ignition of counter space items from an electric range heating element. The two cabinet designs and two ignition scenarios were tested twice. Data collected during the kitchen fire tests included the alarm state of smoke alarms at various locations, the smoke light extinction at three locations at a height of 1.5 m from the floor, and the temperature measurements from thermocouple trees at three locations, and combustion gas sampling at two locations at a height of 1.5 m from the floor. Additionally, carbon dioxide and carbon monoxide were measured in the kitchen at the ceiling location to capture early combustion gases from the fires.

Figure 22 shows the configuration of the kitchen fire tests. The base cabinets mock-ups were constructed from cement board, as well as the two wall cabinets located to the left of the test cabinet. A typical metal range vent hood was installed above the location of the mock-up range, abutting the test cabinet. The range and counter surfaces were covered with aluminum foil to aid with post-test clean up.

Figure 22. Configuration of kitchen counter and cabinet with ignition scenario 1 shown.

3 Results and Analysis

3.1 Smoke Alarm Sensitivity Measurements

The smoke alarm sensitivity measurements provide a reference sensitivity range of different smoke alarm types relative to the cotton smolder smoke experiment. Smoke alarm of the same make and model were placed side-by-side on the ceiling of the FE/DE test section. The positions were labeled front and back. The alarm locations were swapped after three tests, and the average results from each location were computed. While the measuring ionization chamber samples from the centerline of the FE/DE duct, the extinction measurement across the duct is determined by the average of the smoke across the duct at a particular height. A persistent concentration gradient in the duct would tend to bias the results based on location of the smoke alarm.

The results for each smoke alarm type are given in Table 2. Results are provided in terms of MIC current and smoke obscuration (%/ft. per UL reporting and labeling convention).

Alarm	Position	MIC (pA)	Std Dev (pA)	Avg MIC (pA)	Std Dev (pA)	Obsc. (%/ft.)	Std Dev (%/ft.)	Avg Obsc. (%/ft.)	Std Dev (%/ft.)
I1	front	87.2	1.8	87.6	1.8	0.24	0.02	0.22	0.03
	back	87.9	1.9			0.21	0.03		
I2	front	82.4	1.2	81.9	1.4	0.32	.06	0.33	0.06
	back	81.5	1.5			0.34	0.06		
P1	front	52.3	5.6	50.7	4.2	1.42	0.29	1.50	0.23
	back	49.2	1.2			1.58	0.12		
P2	front	54.3	4.3	54.8	3.8	1.33	0.17	1.29	0.14
	back	55.2	3.6			1.26	0.12		
D1	front	73.6	4.3	72.6	3.9	0.54	0.10	0.57	0.13
	back	71.7	3.7			0.60	0.15		
D2	front	80.5	2.1	82.7	2.9	0.35	0.06	0.30	0.08
	back	85.0	1.3			0.24	0.05		
M1	front	65.4	2.2	65.4	1.7	0.78	0.11	0.78	0.09
	back	65.3	1.3			0.78	0.08		
M2	front	82.2	3.6	82.5	2.5	0.30	0.07	0.29	0.05
	back	82.9	1.0			0.28	0.03		

Table 2. Tabulated values of average smoke alarm sensitivity.

It was observed that the difference between the front and back position average MIC current ranged from 0.1 pA to 4.5 pA. In most cases the average front and back MIC current for like alarms fell within the other position's standard deviation. An exception is D2 where the difference between the means was greater than one standard deviation. The average MIC current and obscuration sensitivity including all front and back alarm position results was computed and are listed in the table. The alarm with the highest sensitivity to the cotton smolder smoke is I1, and the alarm with the lowest sensitivity is P1. The relative sensitivities to other smoke sources would vary depending on the smoke characteristics.

3.2   Fire Scenario Heat Release Rates

Each fire scenario and cabinet construction was tested in the NIST furniture calorimeter to determine the heat release rate (HRR) as the fire progressed until it was extinguished, or ceased flaming. In addition to the cabinet constructions and sheet metal barriers, non-combustible cement board cabinet mock-ups were tested to determine the heat release rate of the countertop objects by themselves. Each test was conducted once.

The furniture calorimeter is capable of measuring the heat release rate of furniture-sized objects burning under its exhaust hood. The details the heat release rate calorimetry can be found in reference [13]. The combustion environment in the furniture calorimeter differs from those found in room enclosures. In the free burning conditions of the furniture calorimeter, there is plenty of fresh air entrained into the fire plume. In a room environment, as a fire progresses, the oxygen concentration decreases creating a vitiated environment, typically reducing the burning rate. Combustion in the vitiated room environment leads to increased carbon monoxide concentrations. On the other hand, a hot gas layer that develops in a room environment will radiate heat and tend to increase the burning rate of objects. The furniture calorimeter removes the combustion gases via the exhaust flow, eliminating hot gas layer. Thus, the early fire development in the furniture calorimeter and in a room configuration will tend to match more closely in the early stages of fire than in the later stages. Table 3 gives the measured peak heat release rate and the total heat released for each experiment.

Test Name	Cabinet Construction	Ignition Scenario	Peak Heat Release Rate (kW)	Total Heat Released (MJ)
A1	Oak/Pressboard	1	672	206*
B1	Laminated Pressboard	1	239	65
A1B	Oak/Pressboard, Sheet Metal Barrier	1	111	40
B1B	Laminated Pressboard, Sheet Metal Barrier	1	177	44
A2	Oak/Pressboard	2	107	31
B2	Laminated Pressboard	2	122	29
CB1	Cement Board	1	55	31
CB2	Cement Board	2	59	24

*Fire extinguished approximately 1100 s after ignition.

Table 3. Furniture calorimeter results for the scenarios tested.

The oak/pressboard cabinet ignited with ignition scenario 1 (A1) was essentially completely burned. The peak heat release rate was 2/3 the nominal value required to flashover a small room. Involvement of combustible contents inside the cabinet or spread to an adjacent cabinet could plausibly supply additional fuel to reach flashover. The sheet metal barrier on the oak/pressboard cabinet in test A1B significantly reduced the peak heat release rate and effectively stopped fire propagation to the cabinet. The laminated pressboard cabinet subjected to ignition scenario 1 (B1) experienced the next highest peak heat release rate, and the sheet metal barrier test (B1B) produced a reduced heat release rate. Ignition scenario 2 produced significantly lower peak heat release rates for both cabinet types compared to ignition scenario 1. The cement board tests (CB1 and CB2) reveal the differences in heat release rate between the ignition sources themselves. While the peak heat release rates are similar, the total heat released from CB1 is

21

approximately 30% greater than CB2 which reflects the substantial contribution of the plastic coffee maker to available fuel load.

The heat release rate curve along with a sequence of images showing the fire growth stages are presented in the following figures (23-38). The start time (time =0) of the heat release rate curve was when the power was supplied to the electric hot plate. There was approximately a 100 s elapsed time before the paper towel ignited in each test. Ignition is evident in the initial increase in heat release rate from zero. The picture sequence represents before ignition in the upper left photo, the fire at the peak heat release rate value in the lower left photo, the fire progression at ½ the time to reach the peak heat release rate in the upper right photo, and the end of the test in the lower right photo.

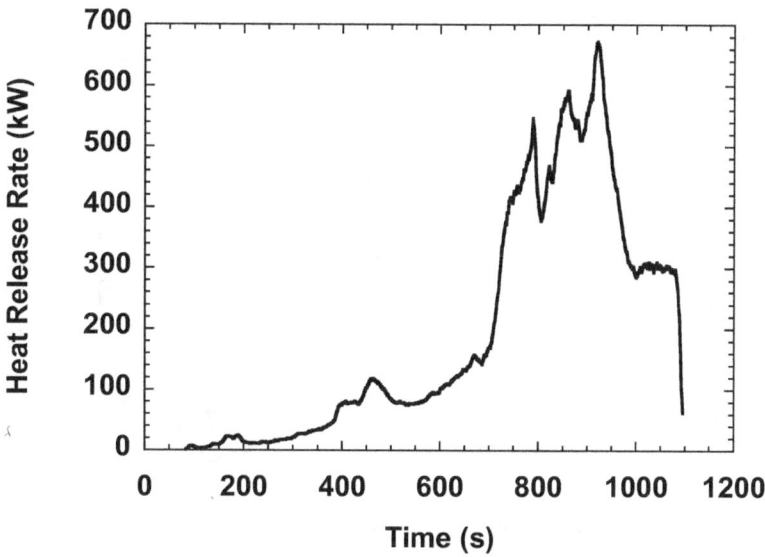

Figure 23. Heat release rate for Test A1 - Oak/pressboard exposed to ignition scenario 1. Fire extinguished with water spray at approximately 1100 s.

Figure 24. Photo sequence for Test A1 - Oak/pressboard exposed to ignition scenario 1.

23

Heat Release Rate (kW)
Time (s)
0
50
100
150
200
250
0
200
400
600
800
1000
1200

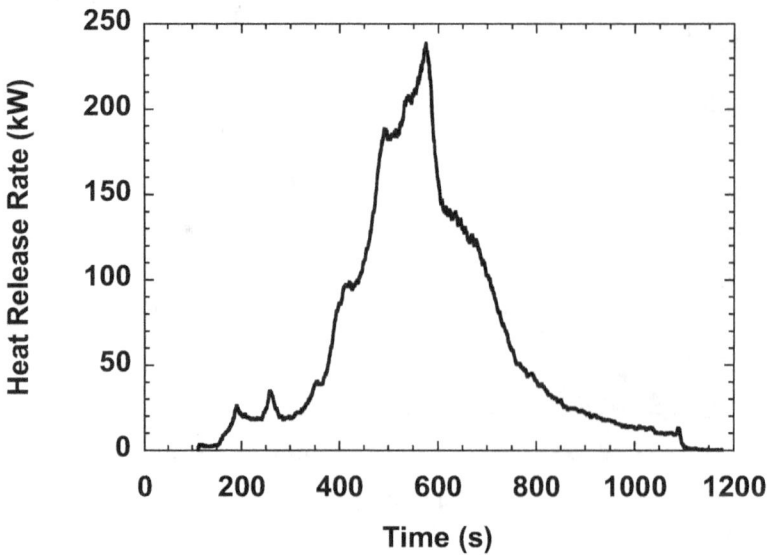

Figure 25. Heat release rate for Test B1 – Laminated pressboard exposed to ignition scenario 1.

SANYO VCC-HD2100: 9/7/2011 13:44:52
SANYO VCC-HD2100: 9/7/2011 13:49:39
SANYO VCC-HD2100: 9/7/2011 13:54:27
SANYO VCC-HD2100: 9/7/2011 14:02:48

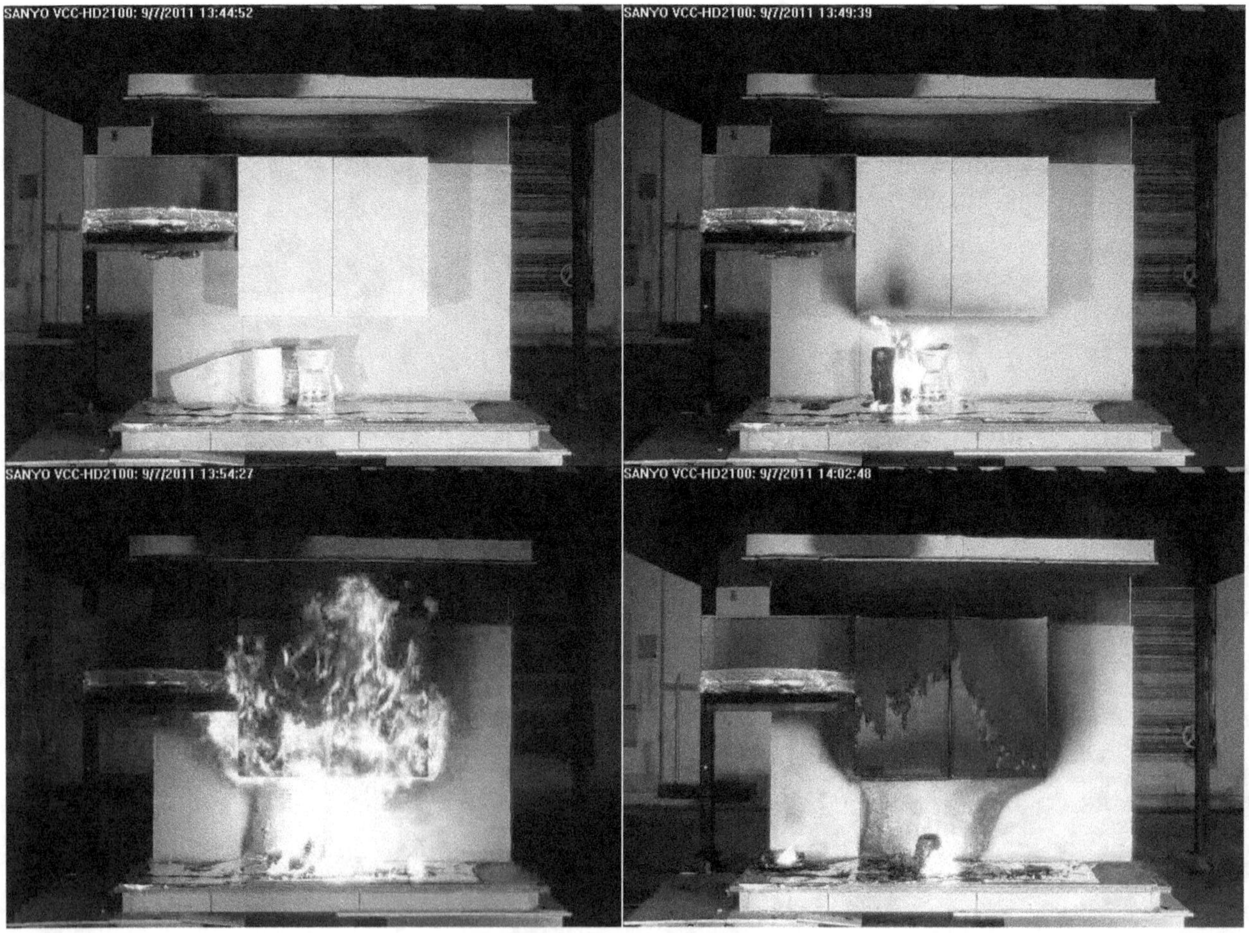

Figure 26. Photo sequence for Test B1 - Laminated pressboard exposed to ignition scenario 1.

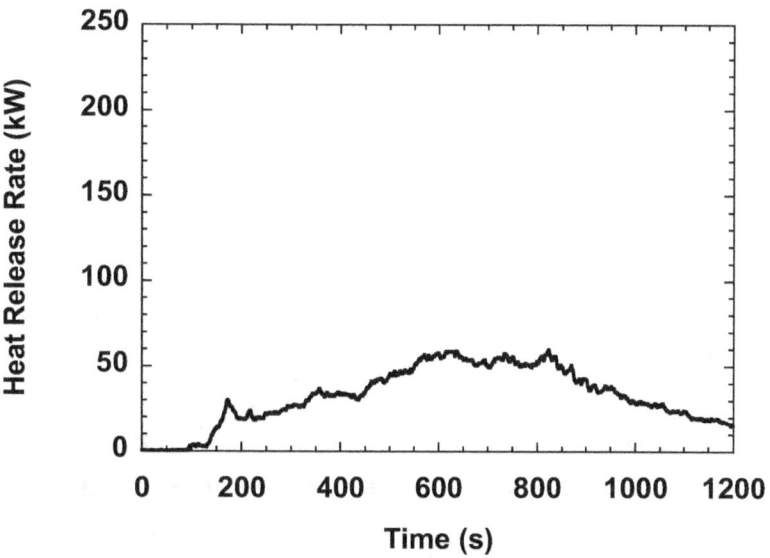

Figure 27. Heat release rate for Test A1B - Oak/pressboard with barrier exposed to ignition scenario 1.

Figure 28. Photo sequence for Test A1B - Oak/pressboard with barrier exposed to ignition scenario

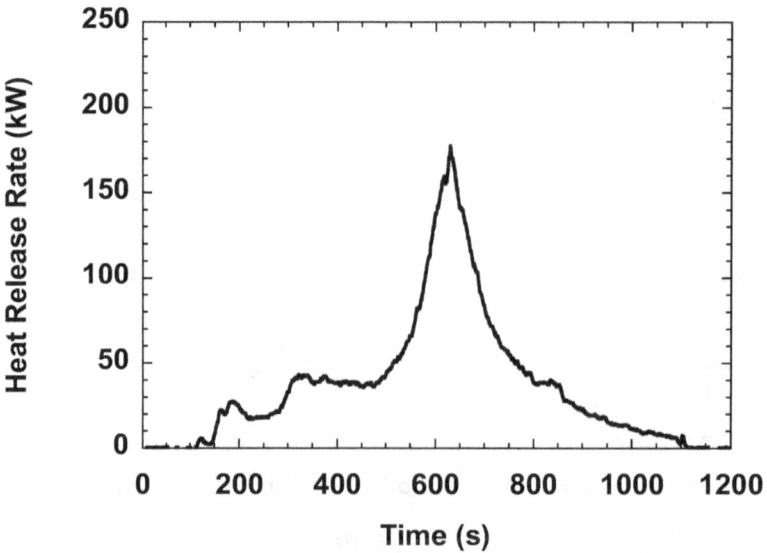

Figure 29. Heat release rate for Test B1B – Laminated pressboard with barrier exposed to ignition scenario 1.

Figure 30. Photo sequence for Test B1B – Laminated pressboard with barrier exposed to ignition scenario 1.

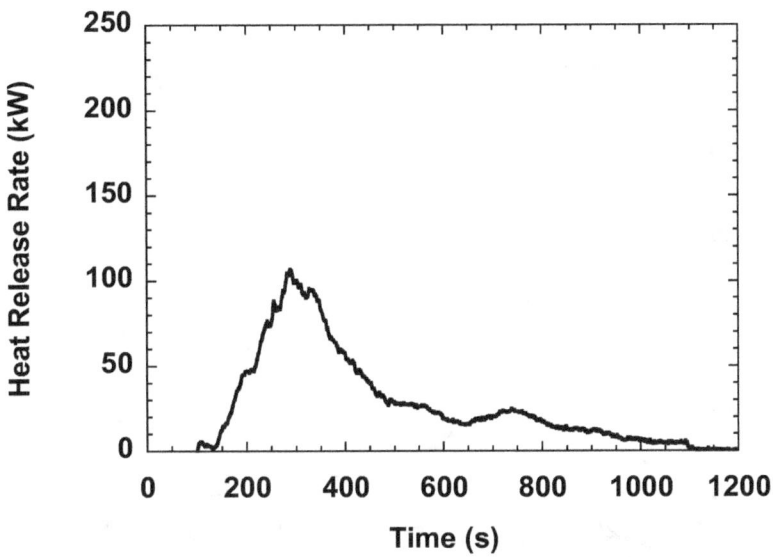

Figure 31. Heat release rate for Test A2 - Oak/pressboard exposed to ignition scenario 2.

Figure 32. Photo sequence for Test A2 - Oak/pressboard exposed to ignition scenario 2.

27

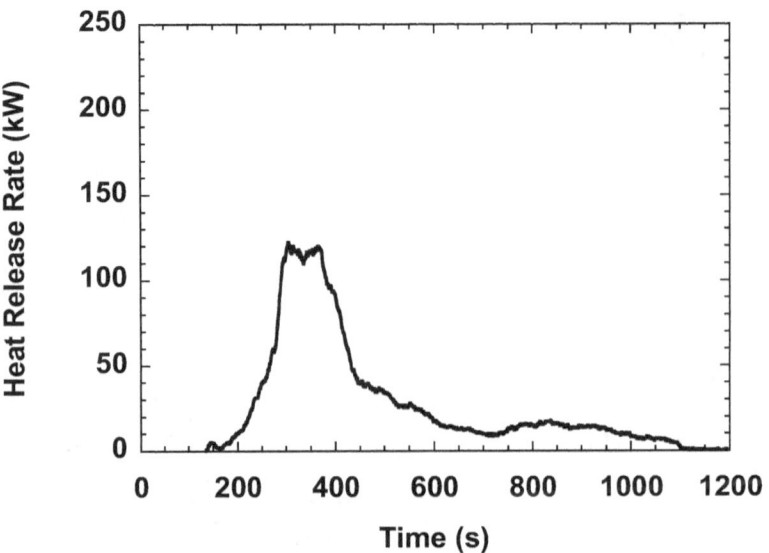

Figure 33. B2 Heat release rate for Test B2 – Laminated pressboard exposed to ignition scenario 2.

![Photo sequence showing four frames of Test B2 fire progression]

Figure 34. Photo sequence for Test B2 - Laminated pressboard exposed to ignition scenario 2.

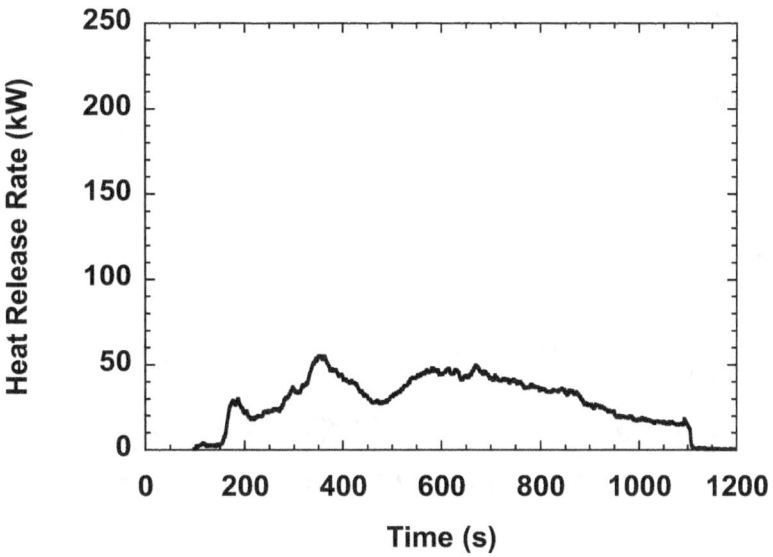

Figure 35. Heat release rate for Test CB1 – Cement board exposed to ignition scenario 1.

Figure 36. Photo sequence for Test CB1 – Cement board exposed to ignition scenario 1.

29

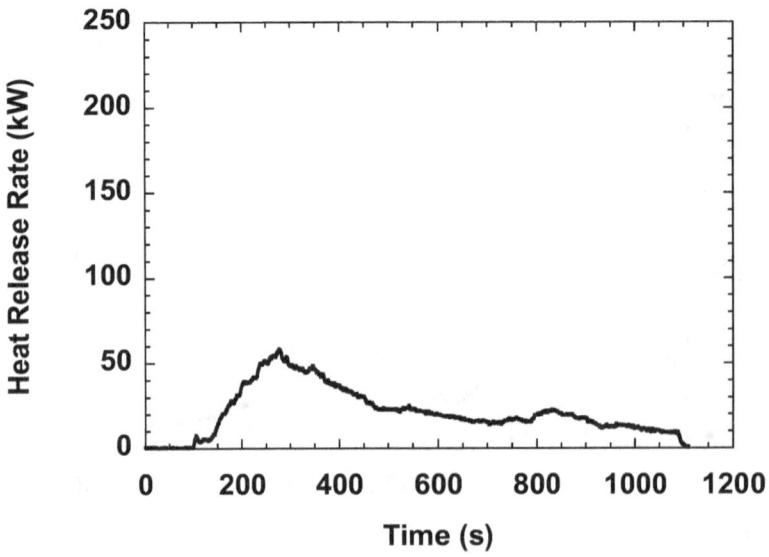

Figure 37. Heat release rate for Test CB2 – Cement board exposed to ignition scenario 2.

![Photo sequence for Test CB2]

Figure 38. Photo sequence for Test CB2 – Cement board exposed to ignition scenario 2.

3.3   Nuisance Alarm Performance

In the burn prop building, there were eight ceiling locations where up to four smoke alarms could be positioned. Two sets of four alarms were mounted on 16 test boards. Every test board contained a P1 and I1 alarm, and the other two alarms were chosen to spread the various types of alarms across the different test boards. One set of alarms was used for the first three tests for each nuisance scenario, and another set of alarms was used for the next three tests for each nuisance alarm scenario. Figure 39 shows the locations of the smoke alarms. Two sets, Loc 1 and Loc 2, were located inside the kitchen at horizontal distances of 1.82 m and 1.87 m from the spot indicated on the range top. Loc 3 and Loc 4 were located outside different kitchen doorways at horizontal distances from the range top of 2.96 m and 3.33 m respectively. Loc 5 - 8 were located in the living room at horizontal distances of 4.50 m, 5.39 m, 6.01 m and 6.94 m, respectively.

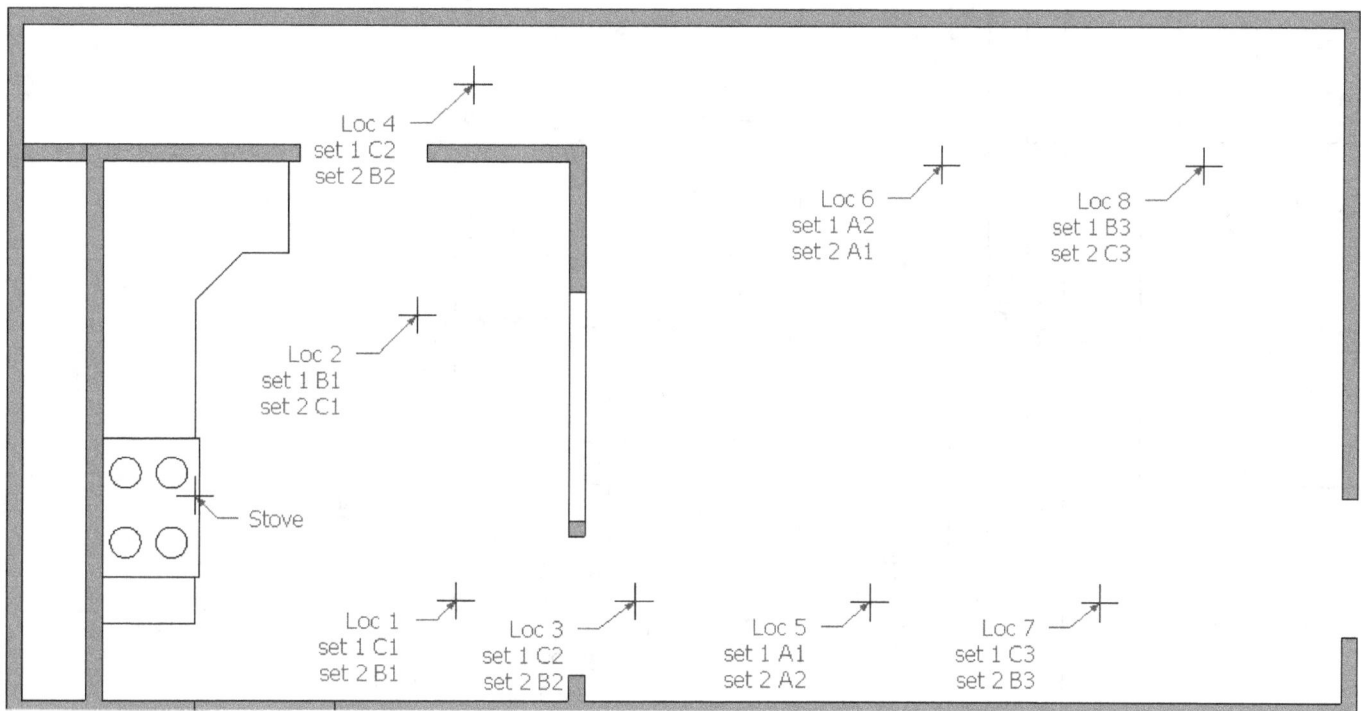

Figure 39. Location of alarms.

Tables 4 to 43 presents the results for the time to alarm for each installed smoke alarm. If the table entry is blank, no alarm was recorded during the test. If the table entry is gray, the particular alarm was not installed during that test.

Distance from Stove, m (ft.)	P1 Exp. 1 T_A (s)	P1 Exp. 2 T_A (s)	P1 Exp. 3 T_A (s)	P1 Exp. 4 T_A (s)	P1 Exp. 5 T_A (s)	P1 Exp. 6 T_A (s)	P1 Alarm Freq.	P2 Exp. 1 T_A (s)	P2 Exp. 2 T_A (s)	P2 Exp. 3 T_A (s)	P2 Exp. 4 T_A (s)	P2 Exp. 5 T_A (s)	P2 Exp. 6 T_A (s)	P2 Alarm Freq.
1.82 (5.98)		375					0.17							NA
1.87 (6.12)	384	384					0.33							NA
2.96 (9.72)		383					0.17							NA
3.33 (10.93)		411					0.17							NA
4.50 (14.77)							0.00							0.00
5.39 (17.70)							0.00							0.00
6.01 (19.71)							0.00							NA
6.94 (22.77)							0.00							NA

Table 4. Photoelectric alarm activation results – frying bacon.

Distance from Stove, m (ft.)	I1 Exp. 1 T_A (s)	I1 Exp. 2 T_A (s)	I1 Exp. 3 T_A (s)	I1 Exp. 4 T_A (s)	I1 Exp. 5 T_A (s)	I1 Exp. 6 T_A (s)	I1 Alarm Freq.	I2 Exp. 1 T_A (s)	I2 Exp. 2 T_A (s)	I2 Exp. 3 T_A (s)	I2 Exp. 4 T_A (s)	I2 Exp. 5 T_A (s)	I2 Exp. 6 T_A (s)	I2 Alarm Freq.
1.82 (5.98)	270	260	272	258	308	307	1.00							NA
1.87 (6.12)	299	278	314	308	334	334	1.00							NA
2.96 (9.72)	317	303	323	294	307	354	1.00							NA
3.33 (10.93)	384	352	320	368	362		0.83							NA
4.50 (14.77)	391	389	363	347	361		0.83	382	383	339	355	371		0.83
5.39 (17.70)	436	397					0.33	445	396	411	457			0.67
6.01 (19.71)		418					0.33				404	412		0.67
6.94 (22.77)							0.00							0.00

Table 5. Ionization alarm activation results – frying bacon.

Table 6. Dual sensor alarm activation results – frying bacon.

Distance from Stove, m (ft.)	D1 Exp. 1 T_A (s)	D1 Exp. 2 T_A (s)	D1 Exp. 3 T_A (s)	D1 Exp. 4 T_A (s)	D1 Exp. 5 T_A (s)	D1 Exp. 6 T_A (s)	D1 Alarm Freq.	D2 Exp. 1 T_A (s)	D2 Exp. 2 T_A (s)	D2 Exp. 3 T_A (s)	D2 Exp. 4 T_A (s)	D2 Exp. 5 T_A (s)	D2 Exp. 6 T_A (s)	D2 Alarm Freq.
1.82 (5.98)				294	315		0.667				259	288	297	1.00
1.87 (6.12)		388					0.333	269	270	274				1.00
2.96 (9.72)				394			0.333				323	337		0.667
3.33 (10.93)		406					0.333	319	330	314				1.00
4.50 (14.77)							NA							NA
5.39 (17.70)							NA							NA
6.01 (19.71)							0.00				370	404		0.667
6.94 (22.77)							0.00							0.00

Table 7. Intelligent alarm activation results – frying bacon.

Distance from Stove, m (ft.)	M1 Exp. 1 T_A (s)	M1 Exp. 2 T_A (s)	M1 Exp. 3 T_A (s)	M1 Exp. 4 T_A (s)	M1 Exp. 5 T_A (s)	M1 Exp. 6 T_A (s)	M1 Alarm Freq.	M2 Exp. 1 T_A (s)	M2 Exp. 2 T_A (s)	M2 Exp. 3 T_A (s)	M2 Exp. 4 T_A (s)	M2 Exp. 5 T_A (s)	M2 Exp. 6 T_A (s)	M2 Alarm Freq.
1.82 (5.98)	357	343	377				1.00	302	290	327				1.00
1.87 (6.12)				314	334	405	1.00							0.00
2.96 (9.72)	380	363					0.67	356	324	364				1.00
3.33 (10.93)							0.00				381			0.33
4.50 (14.77)							NA							NA
5.39 (17.70)							NA							NA
6.01 (19.71)							0.00							0.00
6.94 (22.77)							0.00							0.00

Distance from Stove, m (ft.)	P1 Exp. 1 T_A(s)	P1 Exp. 2 T_A(s)	P1 Exp. 3 T_A(s)	P1 Exp. 4 T_A(s)	P1 Exp. 5 T_A(s)	P1 Exp. 6 T_A(s)	P1 Alarm Freq.	P2 Exp. 1 T_A(s)	P2 Exp. 2 T_A(s)	P2 Exp. 3 T_A(s)	P2 Exp. 4 T_A(s)	P2 Exp. 5 T_A(s)	P2 Exp. 6 T_A(s)	P2 Alarm Freq.
1.82 (5.98)	246		232	232	236	217	0.83							
1.87 (6.12)	248		240		243	215	0.67							
2.96 (9.72)	274		260			241	0.50							
3.33 (10.93)					257	258	0.33							
4.50 (14.77)						257	0.17						251	0.17
5.39 (17.70)							0.00							0.00
6.01 (19.71)						287	0.17							
6.94 (22.77)							0.00							

Table 8. Photoelectric alarm activation results – grilled cheese sandwich.

Distance from Stove, m (ft.)	I1 Exp. 1 T_A(s)	I1 Exp. 2 T_A(s)	I1 Exp. 3 T_A(s)	I1 Exp. 4 T_A(s)	I1 Exp. 5 T_A(s)	I1 Exp. 6 T_A(s)	I1 Alarm Freq.	I2 Exp. 1 T_A(s)	I2 Exp. 2 T_A(s)	I2 Exp. 3 T_A(s)	I2 Exp. 4 T_A(s)	I2 Exp. 5 T_A(s)	I2 Exp. 6 T_A(s)	I2 Alarm Freq.
1.82 (5.98)			232	239		234	0.50							
1.87 (6.12)						243	0.17							
2.96 (9.72)							0.00							
3.33 (10.93)							0.00							
4.50 (14.77)							0.00							0.00
5.39 (17.70)							0.00							0.00
6.01 (19.71)							0.00							0.00
6.94 (22.77)							0.00							0.00

Table 9. Ionization alarm activation results – grilled cheese sandwich.

Table 10. Dual sensor alarm activation results – grilled cheese sandwich.

Distance from Stove, m (ft.)	D1 Exp. 1 T_A (s)	D1 Exp. 2 T_A (s)	D1 Exp. 3 T_A (s)	D1 Exp. 4 T_A (s)	D1 Exp. 5 T_A (s)	D1 Exp. 6 T_A (s)	D1 Alarm Freq.	D2 Exp. 1 T_A (s)	D2 Exp. 2 T_A (s)	D2 Exp. 3 T_A (s)	D2 Exp. 4 T_A (s)	D2 Exp. 5 T_A (s)	D2 Exp. 6 T_A (s)	D2 Alarm Freq.
1.82 (5.98)					234	228	0.67				204	231	223	1.00
1.87 (6.12)			255				0.33			239				0.33
2.96 (9.72)						282	0.33						256	0.33
3.33 (10.93)							0.00							0.00
4.50 (14.77)							NA							NA
5.39 (17.70)							NA							NA
6.01 (19.71)							0.00							0.00
6.94 (22.77)							0.00							0.00

Table 11. Intelligent alarm activation results – grilled cheese sandwich.

Distance from Stove, m (ft.)	M1 Exp. 1 T_A (s)	M1 Exp. 2 T_A (s)	M1 Exp. 3 T_A (s)	M1 Exp. 4 T_A (s)	M1 Exp. 5 T_A (s)	M1 Exp. 6 T_A (s)	M1 Alarm Freq.	M2 Exp. 1 T_A (s)	M2 Exp. 2 T_A (s)	M2 Exp. 3 T_A (s)	M2 Exp. 4 T_A (s)	M2 Exp. 5 T_A (s)	M2 Exp. 6 T_A (s)	M2 Alarm Freq.
1.82 (5.98)			221				0.33							0.00
1.87 (6.12)				119			0.33							0.00
2.96 (9.72)							0.00							0.00
3.33 (10.93)							0.00							0.00
4.50 (14.77)														
5.39 (17.70)														
6.01 (19.71)							0.00							0.00
6.94 (22.77)							0.00							0.00

Distance from Stove, m (ft.)	P1 Exp. 1 T_A (s)	P1 Exp. 2 T_A (s)	P1 Exp. 3 T_A (s)	P1 Exp. 4 T_A (s)	P1 Exp. 5 T_A (s)	P1 Exp. 6 T_A (s)	P1 Alarm Freq.	P2 Exp. 1 T_A (s)	P2 Exp. 2 T_A (s)	P2 Exp. 3 T_A (s)	P2 Exp. 4 T_A (s)	P2 Exp. 5 T_A (s)	P2 Exp. 6 T_A (s)	P2 Alarm Freq.
1.82 (5.98)	296	196	279	321	239	300	1.00							NA
1.87 (6.12)	287	314	290	323	331	317	1.00							NA
2.96 (9.72)	293	241	309	341	316	316	1.00							NA
3.33 (10.93)	330	335	332	379	345	359	1.00							NA
4.50 (14.77)	535		408	394		387	0.67	358	427	403	363	338	343	1.00
5.39 (17.70)							0.00	442	472	437	448	463	433	1.00
6.01 (19.71)					543		0.17							NA
6.94 (22.77)	558		586				0.33							NA

Table 12. Photoelectric alarm activation results – frying hamburger.

Distance from Stove, m (ft.)	I1 Exp. 1 T_A (s)	I1 Exp. 2 T_A (s)	I1 Exp. 3 T_A (s)	I1 Exp. 4 T_A (s)	I1 Exp. 5 T_A (s)	I1 Exp. 6 T_A (s)	I1 Alarm Freq.	I2 Exp. 1 T_A (s)	I2 Exp. 2 T_A (s)	I2 Exp. 3 T_A (s)	I2 Exp. 4 T_A (s)	I2 Exp. 5 T_A (s)	I2 Exp. 6 T_A (s)	I2 Alarm Freq.
1.82 (5.98)	271	266	247	196	236	249	1.00							NA
1.87 (6.12)	282	320	285	202	284	324	1.00							NA
2.96 (9.72)	342	357	307	243	261	275	1.00							NA
3.33 (10.93)	318	279	260	360	344	363	1.00							NA
4.50 (14.77)	489	550	404	360	335	344	1.00	480	550	404	365	332	342	1.00
5.39 (17.70)			555	542			0.33	517	564	409	540	581	576	1.00
6.01 (19.71)							0.00				526	526	409	1.00
6.94 (22.77)	558			544	558	574	1.00							0.00

Table 13. Ionization alarm activation results – frying hamburger.

Distance from Stove, m (ft.)	D1 Exp. 1 T_A(s)	D1 Exp. 2 T_A(s)	D1 Exp. 3 T_A(s)	D1 Exp. 4 T_A(s)	D1 Exp. 5 T_A(s)	D1 Exp. 6 T_A(s)	D1 Alarm Freq.	D2 Exp. 1 T_A(s)	D2 Exp. 2 T_A(s)	D2 Exp. 3 T_A(s)	D2 Exp. 4 T_A(s)	D2 Exp. 5 T_A(s)	D2 Exp. 6 T_A(s)	D2 Alarm Freq.
1.82 (5.98)	315			315	281	278	1.00				172	191	196	1.00
1.87 (6.12)		326	331				1.00	221	210	193				1.00
2.96 (9.72)				337	329	319	1.00				245	300	290	1.00
3.33 (10.93)	377	396	341				1.00	290	330	262				1.00
4.50 (14.77)							NA							NA
5.39 (17.70)							NA							NA
6.01 (19.71)							0.00				439	360	368	1.00
6.94 (22.77)							0.00							0.00

Table 14. Dual sensor alarm activation results – frying hamburger.

Distance from Stove, m (ft.)	M1 Exp. 1 T_A(s)	M1 Exp. 2 T_A(s)	M1 Exp. 3 T_A(s)	M1 Exp. 4 T_A(s)	M1 Exp. 5 T_A(s)	M1 Exp. 6 T_A(s)	M1 Alarm Freq.	M2 Exp. 1 T_A(s)	M2 Exp. 2 T_A(s)	M2 Exp. 3 T_A(s)	M2 Exp. 4 T_A(s)	M2 Exp. 5 T_A(s)	M2 Exp. 6 T_A(s)	M2 Alarm Freq.
1.82 (5.98)	638	504	658				1.00	274	299	275				1.00
1.87 (6.12)				281	369	352	1.00	483	503	488				0.67
2.96 (9.72)			502				0.33				500	502		1.00
3.33 (10.93)							0.00				370	462	366	1.00
4.50 (14.77)							NA							NA
5.39 (17.70)							NA							NA
6.01 (19.71)							0.00							0.00
6.94 (22.77)							0.00							0.00

Table 15. Intelligent alarm activation results – frying hamburger.

Distance from Stove, m (ft.)	P1 Exp. 1 T$_A$ (s)	P1 Exp. 2 T$_A$ (s)	P1 Exp. 3 T$_A$ (s)	P1 Exp. 4 T$_A$ (s)	P1 Exp. 5 T$_A$ (s)	P1 Exp. 6 T$_A$ (s)	P1 Alarm Freq.	P2 Exp. 1 T$_A$ (s)	P2 Exp. 2 T$_A$ (s)	P2 Exp. 3 T$_A$ (s)	P2 Exp. 4 T$_A$ (s)	P2 Exp. 5 T$_A$ (s)	P2 Exp. 6 T$_A$ (s)	P2 Alarm Freq.
1.82 (5.98)	491	448		470		472	0.67							Na
1.87 (6.12)		493	496	464		495	0.67							NA
2.96 (9.72)	479	462	492	493		478	0.83							NA
3.33 (10.93)		508		501		494	0.50							NA
4.50 (14.77)		517		521		515	0.50		499		487		492	0.50
5.39 (17.70)						533	0.17				538		530	0.33
6.01 (19.71)						530	0.17							NA
6.94 (22.77)		601					0.17							NA

Table 16. Photoelectric alarm activation results – stir frying vegetables.

Distance from Stove, m (ft.)	I1 Exp. 1 T$_A$ (s)	I1 Exp. 2 T$_A$ (s)	I1 Exp. 3 T$_A$ (s)	I1 Exp. 4 T$_A$ (s)	I1 Exp. 5 T$_A$ (s)	I1 Exp. 6 T$_A$ (s)	I1 Alarm Freq.	I2 Exp. 1 T$_A$ (s)	I2 Exp. 2 T$_A$ (s)	I2 Exp. 3 T$_A$ (s)	I2 Exp. 4 T$_A$ (s)	I2 Exp. 5 T$_A$ (s)	I2 Exp. 6 T$_A$ (s)	I2 Alarm Freq.
1.82 (5.98)	412	427	470	413	496	453	1.00							NA
1.87 (6.12)	464	468	478	421		492	0.83							NA
2.96 (9.72)	447	460	467	469	516	474	1.00							NA
3.33 (10.93)	499	478	519	502		526	0.83							NA
4.50 (14.77)	493	495		485		509	0.67	483	507		486		508	0.67
5.39 (17.70)							0.00		554					0.17
6.01 (19.71)							0.00							0.00
6.94 (22.77)							0.00							0.00

Table 17. Ionization alarm activation results – stir frying vegetables.

Distance from Stove, m (ft.)	D1 Exp. 1 T$_A$(s)	D1 Exp. 2 T$_A$(s)	D1 Exp. 3 T$_A$(s)	D1 Exp. 4 T$_A$(s)	D1 Exp. 5 T$_A$(s)	D1 Exp. 6 T$_A$(s)	D1 Alarm Freq.	D2 Exp. 1 T$_A$(s)	D2 Exp. 2 T$_A$(s)	D2 Exp. 3 T$_A$(s)	D2 Exp. 4 T$_A$(s)	D2 Exp. 5 T$_A$(s)	D2 Exp. 6 T$_A$(s)	D2 Alarm Freq.
1.82 (5.98)				420		473	0.66				267	460	446	1.00
1.87 (6.12)		489	499				0.66	429	449	475				1.00
2.96 (9.72)						513	0.33				479	519	475	1.00
3.33 (10.93)		520					0.33	485	470	508				1.00
4.50 (14.77)							NA							NA
5.39 (17.70)							NA							NA
6.01 (19.71)							0.00				484		515	0.66
6.94 (22.77)							0.00							0.00

Table 18. Dual sensor alarm activation results – stir frying vegetables.

Distance from Stove, m (ft.)	M1 Exp. 1 T$_A$(s)	M1 Exp. 2 T$_A$(s)	M1 Exp. 3 T$_A$(s)	M1 Exp. 4 T$_A$(s)	M1 Exp. 5 T$_A$(s)	M1 Exp. 6 T$_A$(s)	M1 Alarm Freq.	M2 Exp. 1 T$_A$(s)	M2 Exp. 2 T$_A$(s)	M2 Exp. 3 T$_A$(s)	M2 Exp. 4 T$_A$(s)	M2 Exp. 5 T$_A$(s)	M2 Exp. 6 T$_A$(s)	M2 Alarm Freq.
1.82 (5.98)	497						0.33	451	437	486				1.00
1.87 (6.12)				459	546	580	1.00							0.00
2.96 (9.72)							0.00		495					0.33
3.33 (10.93)							0.00				506			0.33
4.50 (14.77)							NA							NA
5.39 (17.70)							NA							NA
6.01 (19.71)							0.00							0.00
6.94 (22.77)							0.00							0.00

Table 19. Intelligent alarm activation results – stir frying vegetables.

Distance from Stove, m (ft.)	P1 Exp. 1 T$_A$ (s)	P1 Exp. 2 T$_A$ (s)	P1 Exp. 3 T$_A$ (s)	P1 Exp. 4 T$_A$ (s)	P1 Exp. 5 T$_A$ (s)	P1 Exp. 6 T$_A$ (s)	P1 Alarm Freq.	P2 Exp. 1 T$_A$ (s)	P2 Exp. 2 T$_A$ (s)	P2 Exp. 3 T$_A$ (s)	P2 Exp. 4 T$_A$ (s)	P2 Exp. 5 T$_A$ (s)	P2 Exp. 6 T$_A$ (s)	P2 Alarm Freq.
1.82 (5.98)	617	615					0.33							NA
1.87 (6.12)	625	616					0.33							NA
2.96 (9.72)	638	640					0.33							NA
3.33 (10.93)							0.00							NA
4.50 (14.77)							0.00							0.00
5.39 (17.70)							0.00							0.00
6.01 (19.71)							0.00							NA
6.94 (22.77)							0.00							NA

Table 20. Photoelectric alarm activation results – broiling hamburger.

Distance from Stove, m (ft.)	I1 Exp. 1 T$_A$ (s)	I1 Exp. 2 T$_A$ (s)	I1 Exp. 3 T$_A$ (s)	I1 Exp. 4 T$_A$ (s)	I1 Exp. 5 T$_A$ (s)	I1 Exp. 6 T$_A$ (s)	I1 Alarm Freq.	I2 Exp. 1 T$_A$ (s)	I2 Exp. 2 T$_A$ (s)	I2 Exp. 3 T$_A$ (s)	I2 Exp. 4 T$_A$ (s)	I2 Exp. 5 T$_A$ (s)	I2 Exp. 6 T$_A$ (s)	I2 Alarm Freq.
1.82 (5.98)	549	543	539	534	500	513	1.00							NA
1.87 (6.12)	534	505	507	553	506	576	1.00							NA
2.96 (9.72)	548	539	518	551	532	554	1.00							NA
3.33 (10.93)	554	531	496	606	573	613	1.00							NA
4.50 (14.77)	581	588	566	629	583	711	1.00	579	560	536	619	569	561	1.00
5.39 (17.70)	652	645	674	852	770	875	1.00	610	617	621	660	655	734	1.00
6.01 (19.71)	716	651	646				1.00				640	617	650	1.00
6.94 (22.77)				679	669	705	1.00	790	668	727				1.00

Table 21. Ionization alarm activation results – broiling hamburger.

Table 22. Dual sensor alarm activation results – broiling hamburger.

Distance from Stove, m (ft.)	D1 Exp. 1 T_A (s)	D1 Exp. 2 T_A (s)	D1 Exp. 3 T_A (s)	D1 Exp. 4 T_A (s)	D1 Exp. 5 T_A (s)	D1 Exp. 6 T_A (s)	D1 Alarm Freq.	D2 Exp. 1 T_A (s)	D2 Exp. 2 T_A (s)	D2 Exp. 3 T_A (s)	D2 Exp. 4 T_A (s)	D2 Exp. 5 T_A (s)	D2 Exp. 6 T_A (s)	D2 Alarm Freq.
1.82 (5.98)				613	584	700	1.00				466	463	442	1.00
1.87 (6.12)	869	757					0.67	522	515	508				1.00
2.96 (9.72)				801			0.33				583	574	580	1.00
3.33 (10.93)	626	613	631				1.00	522	501	595				1.00
4.50 (14.77)							NA							NA
5.39 (17.70)							NA							NA
6.01 (19.71)							0.00				590	560	606	1.00
6.94 (22.77)							0.00	646	631	653				1.00

Table 23. Intelligent alarm activation results – broiling hamburger.

Distance from Stove, m (ft.)	M1 Exp. 1 T_A (s)	M1 Exp. 2 T_A (s)	M1 Exp. 3 T_A (s)	M1 Exp. 4 T_A (s)	M1 Exp. 5 T_A (s)	M1 Exp. 6 T_A (s)	M1 Alarm Freq.	M2 Exp. 1 T_A (s)	M2 Exp. 2 T_A (s)	M2 Exp. 3 T_A (s)	M2 Exp. 4 T_A (s)	M2 Exp. 5 T_A (s)	M2 Exp. 6 T_A (s)	M2 Alarm Freq.
1.82 (5.98)	558	501	741				1.00	574	579	581				1.00
1.87 (6.12)				506	525	360	1.00					841		0.33
2.96 (9.72)	811	795	918				1.00	757	621					0.67
3.33 (10.93)				725	951	950	1.00				810	673	812	1.00
4.50 (14.77)							NA							NA
5.39 (17.70)							NA							NA
6.01 (19.71)	553	922	879				1.00	868	881					0.67
6.94 (22.77)							0.00							0.00

Distance from Stove, m (ft.)	P1 Exp. 1 T_A(s)	P1 Exp. 2 T_A(s)	P1 Exp. 3 T_A(s)	P1 Exp. 4 T_A(s)	P1 Exp. 5 T_A(s)	P1 Exp. 6 T_A(s)	P1 Alarm Freq.	P2 Exp. 1 T_A(s)	P2 Exp. 2 T_A(s)	P2 Exp. 3 T_A(s)	P2 Exp. 4 T_A(s)	P2 Exp. 5 T_A(s)	P2 Exp. 6 T_A(s)	P2 Alarm Freq.
1.82 (5.98)							0.00							NA
1.87 (6.12)							0.00							NA
2.96 (9.72)							0.00							NA
3.33 (10.93)							0.00							NA
4.50 (14.77)							0.00							0.00
5.39 (17.70)							0.00							0.00
6.01 (19.71)							0.00							NA
6.94 (22.77)							0.00							NA

Table 24. Photoelectric alarm activation results – baking pizza.

Distance from Stove, m (ft.)	I1 Exp. 1 T_A(s)	I1 Exp. 2 T_A(s)	I1 Exp. 3 T_A(s)	I1 Exp. 4 T_A(s)	I1 Exp. 5 T_A(s)	I1 Exp. 6 T_A(s)	I1 Alarm Freq.	I2 Exp. 1 T_A(s)	I2 Exp. 2 T_A(s)	I2 Exp. 3 T_A(s)	I2 Exp. 4 T_A(s)	I2 Exp. 5 T_A(s)	I2 Exp. 6 T_A(s)	I2 Alarm Freq.
1.82 (5.98)	608	608	9	8	632	609	0.50							NA
1.87 (6.12)				7	16		0.83							NA
2.96 (9.72)							0.00							NA
3.33 (10.93)							0.00							NA
4.50 (14.77)							0.00							0.00
5.39 (17.70)							0.00							0.00
6.01 (19.71)							0.00							0.00
6.94 (22.77)							0.00							0.00

Table 25. Ionization alarm activation results – baking pizza.

Distance from Stove, m (ft.)	D1 Exp. 1 T_A (s)	D1 Exp. 2 T_A (s)	D1 Exp. 3 T_A (s)	D1 Exp. 4 T_A (s)	D1 Exp. 5 T_A (s)	D1 Exp. 6 T_A (s)	D1 Alarm Freq.	D2 Exp. 1 T_A (s)	D2 Exp. 2 T_A (s)	D2 Exp. 3 T_A (s)	D2 Exp. 4 T_A (s)	D2 Exp. 5 T_A (s)	D2 Exp. 6 T_A (s)	D2 Alarm Freq.
1.82 (5.98)							0.00				10	11	12	1.00
1.87 (6.12)							0.00			12				0.33
2.96 (9.72)							0.00				25			0.33
3.33 (10.93)							0.00							0.00
4.50 (14.77)							NA							NA
5.39 (17.70)							NA							NA
6.01 (19.71)							0.00				631			0.33
6.94 (22.77)							0.00							0.00

Table 26. Dual sensor alarm activation results – baking pizza.

Distance from Stove, m (ft.)	M1 Exp. 1 T_A (s)	M1 Exp. 2 T_A (s)	M1 Exp. 3 T_A (s)	M1 Exp. 4 T_A (s)	M1 Exp. 5 T_A (s)	M1 Exp. 6 T_A (s)	M1 Alarm Freq.	M2 Exp. 1 T_A (s)	M2 Exp. 2 T_A (s)	M2 Exp. 3 T_A (s)	M2 Exp. 4 T_A (s)	M2 Exp. 5 T_A (s)	M2 Exp. 6 T_A (s)	M2 Alarm Freq.
1.82 (5.98)	265	4	264				1.00							0.00
1.87 (6.12)				39	456	500	1.00							0.00
2.96 (9.72)	262	287	545				1.00							0.00
3.33 (10.93)				176			0.33							0.00
4.50 (14.77)							NA							NA
5.39 (17.70)							NA							NA
6.01 (19.71)	448	434	458				1.00							0.00
6.94 (22.77)							0.00							0.00

Table 27. Intelligent alarm activation results – baking pizza.

Distance from Stove, m (ft.)	P1 Exp. 1 T_A (s)	P1 Exp. 2 T_A (s)	P1 Exp. 3 T_A (s)	P1 Exp. 4 T_A (s)	P1 Exp. 5 T_A (s)	P1 Exp. 6 T_A (s)	P1 Alarm Freq.	P2 Exp. 1 T_A (s)	P2 Exp. 2 T_A (s)	P2 Exp. 3 T_A (s)	P2 Exp. 4 T_A (s)	P2 Exp. 5 T_A (s)	P2 Exp. 6 T_A (s)	P2 Alarm Freq.
1.82 (5.98)							0.00							NA
1.87 (6.12)							0.00							NA
2.96 (9.72)							0.00							NA
3.33 (10.93)							0.00							NA
4.50 (14.77)							0.00							0.00
5.39 (17.70)							0.00							0.00
6.01 (19.71)							0.00							NA
6.94 (22.77)							0.00							NA

Table 28. Photoelectric alarm activation results – light toast.

Distance from Stove, m (ft.)	I1 Exp. 1 T_A (s)	I1 Exp. 2 T_A (s)	I1 Exp. 3 T_A (s)	I1 Exp. 4 T_A (s)	I1 Exp. 5 T_A (s)	I1 Exp. 6 T_A (s)	I1 Alarm Freq.	I2 Exp. 1 T_A (s)	I2 Exp. 2 T_A (s)	I2 Exp. 3 T_A (s)	I2 Exp. 4 T_A (s)	I2 Exp. 5 T_A (s)	I2 Exp. 6 T_A (s)	I2 Alarm Freq.
1.82 (5.98)							0.00							NA
1.87 (6.12)	113						0.17							NA
2.96 (9.72)							0.00							NA
3.33 (10.93)							0.00							NA
4.50 (14.77)							0.00							0.00
5.39 (17.70)							0.00							0.00
6.01 (19.71)							0.00							0.00
6.94 (22.77)							0.00							0.00

Table 29. Ionization alarm activation results – light toast.

44

Distance from Stove, m (ft.)	D1 Exp. 1 T_A (s)	D1 Exp. 2 T_A (s)	D1 Exp. 3 T_A (s)	D1 Exp. 4 T_A (s)	D1 Exp. 5 T_A (s)	D1 Exp. 6 T_A (s)	D1 Alarm Freq.	D2 Exp. 1 T_A (s)	D2 Exp. 2 T_A (s)	D2 Exp. 3 T_A (s)	D2 Exp. 4 T_A (s)	D2 Exp. 5 T_A (s)	D2 Exp. 6 T_A (s)	D2 Alarm Freq.
1.82 (5.98)							0.00							0.00
1.87 (6.12)							0.00							0.00
2.96 (9.72)							0.00							0.00
3.33 (10.93)							0.00							0.00
4.50 (14.77)							NA							NA
5.39 (17.70)							NA							NA
6.01 (19.71)							0.00							0.00
6.94 (22.77)							0.00							0.00

Table 30. Dual sensor alarm activation results – light toast.

Distance from Stove, m (ft.)	M1 Exp. 1 T_A (s)	M1 Exp. 2 T_A (s)	M1 Exp. 3 T_A (s)	M1 Exp. 4 T_A (s)	M1 Exp. 5 T_A (s)	M1 Exp. 6 T_A (s)	M1 Alarm Freq.	M2 Exp. 1 T_A (s)	M2 Exp. 2 T_A (s)	M2 Exp. 3 T_A (s)	M2 Exp. 4 T_A (s)	M2 Exp. 5 T_A (s)	M2 Exp. 6 T_A (s)	M2 Alarm Freq.
1.82 (5.98)							0.00							0.00
1.87 (6.12)							0.00							0.00
2.96 (9.72)							0.00							0.00
3.33 (10.93)							0.00							0.00
4.50 (14.77)							NA							NA
5.39 (17.70)							NA							NA
6.01 (19.71)							0.00							0.00
6.94 (22.77)							0.00							0.00

Table 31. Intelligent alarm activation results – light toast.

Distance from Stove, m (ft.)	P1 Exp. 1 T$_A$(s)	P1 Exp. 2 T$_A$(s)	P1 Exp. 3 T$_A$(s)	P1 Exp. 4 T$_A$(s)	P1 Exp. 5 T$_A$(s)	P1 Exp. 6 T$_A$(s)	P1 Alarm Freq.	P2 Exp. 1 T$_A$(s)	P2 Exp. 2 T$_A$(s)	P2 Exp. 3 T$_A$(s)	P2 Exp. 4 T$_A$(s)	P2 Exp. 5 T$_A$(s)	P2 Exp. 6 T$_A$(s)	P2 Alarm Freq.
1.82 (5.98)							0.00							NA
1.87 (6.12)							0.00							NA
2.96 (9.72)							0.00							NA
3.33 (10.93)							0.00							NA
4.50 (14.77)							0.00							0.00
5.39 (17.70)							0.00							0.00
6.01 (19.71)							0.00							NA
6.94 (22.77)							0.00							NA

Table 32. Photoelectric alarm activation results – dark toast.

Distance from Stove, m (ft.)	I1 Exp. 1 T$_A$(s)	I1 Exp. 2 T$_A$(s)	I1 Exp. 3 T$_A$(s)	I1 Exp. 4 T$_A$(s)	I1 Exp. 5 T$_A$(s)	I1 Exp. 6 T$_A$(s)	I1 Alarm Freq.	I2 Exp. 1 T$_A$(s)	I2 Exp. 2 T$_A$(s)	I2 Exp. 3 T$_A$(s)	I2 Exp. 4 T$_A$(s)	I2 Exp. 5 T$_A$(s)	I2 Exp. 6 T$_A$(s)	I2 Alarm Freq.
1.82 (5.98)	164	161	155	192	171	167	1.00							NA
1.87 (6.12)	116	121	129	147	137	156	1.00							NA
2.96 (9.72)	221	237	215	232	204	209	1.00							NA
3.33 (10.93)							0.00							NA
4.50 (14.77)							0.00							0.00
5.39 (17.70)							0.00							0.00
6.01 (19.71)							0.00							0.00
6.94 (22.77)							0.00							0.00

Table 33. Ionization alarm activation results – dark toast.

Distance from Stove, m (ft.)	D1 Exp. 1 T_A (s)	D1 Exp. 2 T_A (s)	D1 Exp. 3 T_A (s)	D1 Exp. 4 T_A (s)	D1 Exp. 5 T_A (s)	D1 Exp. 6 T_A (s)	D1 Alarm Freq.	D2 Exp. 1 T_A (s)	D2 Exp. 2 T_A (s)	D2 Exp. 3 T_A (s)	D2 Exp. 4 T_A (s)	D2 Exp. 5 T_A (s)	D2 Exp. 6 T_A (s)	D2 Alarm Freq.
1.82 (5.98)					183	185	0.67				157	174	171	1.00
1.87 (6.12)	173	191	192				1.00	173	191	192				1.00
2.96 (9.72)							0.00							0.00
3.33 (10.93)							0.00							0.00
4.50 (14.77)							NA							NA
5.39 (17.70)							NA							NA
6.01 (19.71)							0.00							0.00
6.94 (22.77)							0.00							0.00

Table 34. Dual sensor alarm activation results – dark toast.

Distance from Stove, m (ft.)	M1 Exp. 1 T_A (s)	M1 Exp. 2 T_A (s)	M1 Exp. 3 T_A (s)	M1 Exp. 4 T_A (s)	M1 Exp. 5 T_A (s)	M1 Exp. 6 T_A (s)	M1 Alarm Freq.	M2 Exp. 1 T_A (s)	M2 Exp. 2 T_A (s)	M2 Exp. 3 T_A (s)	M2 Exp. 4 T_A (s)	M2 Exp. 5 T_A (s)	M2 Exp. 6 T_A (s)	M2 Alarm Freq.
1.82 (5.98)	226						0.33				177	186	185	1.00
1.87 (6.12)					179	169	0.67	190	196	179				1.00
2.96 (9.72)							0.00							0.00
3.33 (10.93)							0.00							0.00
4.50 (14.77)							NA							NA
5.39 (17.70)							NA							NA
6.01 (19.71)							0.00							0.00
6.94 (22.77)							0.00							0.00

Table 35. Intelligent alarm activation results – dark toast.

Distance from Stove, m (ft.)	P1 Exp. 1 T$_A$(s)	P1 Exp. 2 T$_A$(s)	P1 Exp. 3 T$_A$(s)	P1 Exp. 4 T$_A$(s)	P1 Exp. 5 T$_A$(s)	P1 Exp. 6 T$_A$(s)	P1 Alarm Freq.	P2 Exp. 1 T$_A$(s)	P2 Exp. 2 T$_A$(s)	P2 Exp. 3 T$_A$(s)	P2 Exp. 4 T$_A$(s)	P2 Exp. 5 T$_A$(s)	P2 Exp. 6 T$_A$(s)	P2 Alarm Freq.
1.82 (5.98)	359	352	349	368	349	346	1.00							NA
1.87 (6.12)	332	336	323	333	328	346	1.00							NA
2.96 (9.72)	366	356	351	370	367	381	1.00							NA
3.33 (10.93)	388	414	384	446	438	419	1.00							NA
4.50 (14.77)	411	497	396	427	517	565	1.00	412	430	389	409	423	414	1.00
5.39 (17.70)	489		465				0.33	481		446				0.33
6.01 (19.71)					594		0.17							NA
6.94 (22.77)							0.00							NA

Table 36. Photoelectric alarm activation results – very dark toast.

Distance from Stove, m (ft.)	I1 Exp. 1 T$_A$(s)	I1 Exp. 2 T$_A$(s)	I1 Exp. 3 T$_A$(s)	I1 Exp. 4 T$_A$(s)	I1 Exp. 5 T$_A$(s)	I1 Exp. 6 T$_A$(s)	I1 Alarm Freq.	I2 Exp. 1 T$_A$(s)	I2 Exp. 2 T$_A$(s)	I2 Exp. 3 T$_A$(s)	I2 Exp. 4 T$_A$(s)	I2 Exp. 5 T$_A$(s)	I2 Exp. 6 T$_A$(s)	I2 Alarm Freq.
1.82 (5.98)	182	188	194	196	183	187	1.00							NA
1.87 (6.12)	152	150	160	165	161	155	1.00							NA
2.96 (9.72)	231	209	214	216	216	241	1.00							NA
3.33 (10.93)	439	347	375	257	298	295	1.00							NA
4.50 (14.77)	316	272	320	361	314	341	1.00	315	281	334	323	314	341	1.00
5.39 (17.70)				424	424		0.33				474			0.17
6.01 (19.71)			332				0.33					624		0.33
6.94 (22.77)							0.00							0.00

Table 37. Ionization alarm activation results – very dark toast.

Table 38. Dual sensor alarm activation results.

Distance from Stove, m (ft.)	D1 Exp. 1 T_A (s)	D1 Exp. 2 T_A (s)	D1 Exp. 3 T_A (s)	D1 Exp. 4 T_A (s)	D1 Exp. 5 T_A (s)	D1 Exp. 6 T_A (s)	D1 Alarm Freq.	D2 Exp. 1 T_A (s)	D2 Exp. 2 T_A (s)	D2 Exp. 3 T_A (s)	D2 Exp. 4 T_A (s)	D2 Exp. 5 T_A (s)	D2 Exp. 6 T_A (s)	D2 Alarm Freq.
1.82 (5.98)	196	206	206				1.00	180	173	177				1.00
1.87 (6.12)				227	218	237	1.00				189	184	173	1.00
2.96 (9.72)	418	385	397				1.00	292	250	263				1.00
3.33 (10.93)				267	266	303	1.00				269	245	235	1.00
4.50 (14.77)							NA							NA
5.39 (17.70)							NA							NA
6.01 (19.71)						624	0.33					397	372	0.67
6.94 (22.77)							0.00							0.00

Table 38. Dual sensor alarm activation results – very dark toast.

Distance from Stove, m (ft.)	M1 Exp. 1 T_A (s)	M1 Exp. 2 T_A (s)	M1 Exp. 3 T_A (s)	M1 Exp. 4 T_A (s)	M1 Exp. 5 T_A (s)	M1 Exp. 6 T_A (s)	M1 Alarm Freq.	M2 Exp. 1 T_A (s)	M2 Exp. 2 T_A (s)	M2 Exp. 3 T_A (s)	M2 Exp. 4 T_A (s)	M2 Exp. 5 T_A (s)	M2 Exp. 6 T_A (s)	M2 Alarm Freq.
1.82 (5.98)	248	236	253				1.00	211	200	194				1.00
1.87 (6.12)				202	191	205	1.00				201	210	204	1.00
2.96 (9.72)	288	268	294				1.00	366	356	351				1.00
3.33 (10.93)				317	382		0.67				329	344	363	1.00
4.50 (14.77)														
5.39 (17.70)														
6.01 (19.71)							0.00							0.00
6.94 (22.77)							0.00							0.00

Table 39. Intelligent alarm activation results – very dark toast.

Distance from Stove, m (ft.)	P1 Exp. 1 T$_A$(s)	P1 Exp. 2 T$_A$(s)	P1 Exp. 3 T$_A$(s)	P1 Exp. 4 T$_A$(s)	P1 Exp. 5 T$_A$(s)	P1 Exp. 6 T$_A$(s)	P1 Alarm Freq.	P2 Exp. 1 T$_A$(s)	P2 Exp. 2 T$_A$(s)	P2 Exp. 3 T$_A$(s)	P2 Exp. 4 T$_A$(s)	P2 Exp. 5 T$_A$(s)	P2 Exp. 6 T$_A$(s)	P2 Alarm Freq.
1.82 (5.98)							0.00							NA
1.87 (6.12)							0.00							NA
2.96 (9.72)							0.00							NA
3.33 (10.93)							0.00							NA
4.50 (14.77)							0.00							0.00
5.39 (17.70)							0.00							0.00
6.01 (19.71)							0.00							NA
6.94 (22.77)							0.00							NA

Table 40. Photoelectric alarm activation results – toasting bagel.

Distance from Stove, m (ft.)	I1 Exp. 1 T$_A$(s)	I1 Exp. 2 T$_A$(s)	I1 Exp. 3 T$_A$(s)	I1 Exp. 4 T$_A$(s)	I1 Exp. 5 T$_A$(s)	I1 Exp. 6 T$_A$(s)	I1 Alarm Freq.	I2 Exp. 1 T$_A$(s)	I2 Exp. 2 T$_A$(s)	I2 Exp. 3 T$_A$(s)	I2 Exp. 4 T$_A$(s)	I2 Exp. 5 T$_A$(s)	I2 Exp. 6 T$_A$(s)	I2 Alarm Freq.
1.82 (5.98)	197	201	160	192	178	186	1.00							NA
1.87 (6.12)	149	153	138	178	155	160	1.00							NA
2.96 (9.72)		237	233			229	0.50							NA
3.33 (10.93)							0.00							NA
4.50 (14.77)							0.00							0.00
5.39 (17.70)							0.00							0.00
6.01 (19.71)							0.00							0.00
6.94 (22.77)							0.00							0.00

Table 41. Ionization alarm activation results – toasting bagel.

Table 42.

Distance from Stove, m (ft.)	D1 Exp. 1 T$_A$(s)	D1 Exp. 2 T$_A$(s)	D1 Exp. 3 T$_A$(s)	D1 Exp. 4 T$_A$(s)	D1 Exp. 5 T$_A$(s)	D1 Exp. 6 T$_A$(s)	D1 Alarm Freq.	D2 Exp. 1 T$_A$(s)	D2 Exp. 2 T$_A$(s)	D2 Exp. 3 T$_A$(s)	D2 Exp. 4 T$_A$(s)	D2 Exp. 5 T$_A$(s)	D2 Exp. 6 T$_A$(s)	D2 Alarm Freq.
1.82 (5.98)				226		194	0.67				182	193	184	1.00
1.87 (6.12)		218	213				0.67	171	173	160				1.00
2.96 (9.72)							0.00				256	240	230	1.00
3.33 (10.93)							0.00							0.00
4.50 (14.77)							NA							NA
5.39 (17.70)							NA							NA
6.01 (19.71)							0.00							0.00
6.94 (22.77)							0.00							0.00

Table 42. Dual sensor alarm activation results – toasting bagel.

Table 43.

Distance from Stove, m (ft.)	M1 Exp. 1 T$_A$(s)	M1 Exp. 2 T$_A$(s)	M1 Exp. 3 T$_A$(s)	M1 Exp. 4 T$_A$(s)	M1 Exp. 5 T$_A$(s)	M1 Exp. 6 T$_A$(s)	M1 Alarm Freq.	M2 Exp. 1 T$_A$(s)	M2 Exp. 2 T$_A$(s)	M2 Exp. 3 T$_A$(s)	M2 Exp. 4 T$_A$(s)	M2 Exp. 5 T$_A$(s)	M2 Exp. 6 T$_A$(s)	M2 Alarm Freq.
1.82 (5.98)							0.00	214		219				0.67
1.87 (6.12)							0.00					212	206	0.67
2.96 (9.72)							0.00							0.00
3.33 (10.93)							0.00							0.00
4.50 (14.77)							NA							NA
5.39 (17.70)							NA							NA
6.01 (19.71)							0.00							0.00
6.94 (22.77)							0.00							0.00

Table 43. Intelligent alarm activation results – toasting bagel.

The propensity of an alarm to activate appears to be a function of the type of alarm, its sensitivity, its distance from the cooking activity, and the cooking event itself. For example, only one ionization alarm activated during the six light toasting experiments, while most alarms within 4.5 m of the range activated during the six very dark toast experiments. In order to analyze alarm activation tendencies, the results from similar cooking activities were aggregated, as were the results for alarm location pairs 1-2, 3-4, 5-6, and 7-8. Results from cooking activities that used the electric range, oven, or toaster were aggregated for individual alarms and for alarm locations. The distances from the cooking source to the alarm location pairs were averaged to present the results as a function of distance.

Figure 40 shows the fraction of specific types of alarms activated during the aggregated range top cooking events as a function of distance. In general, the fraction of activated alarms decreased as the distance from the cooking source increased, as expected. D2 appears to be the most sensitive to nuisance alarm during the range top cooking activities, while D1, M1, and M2 all had no activations at the farthest distance from the cooking source.

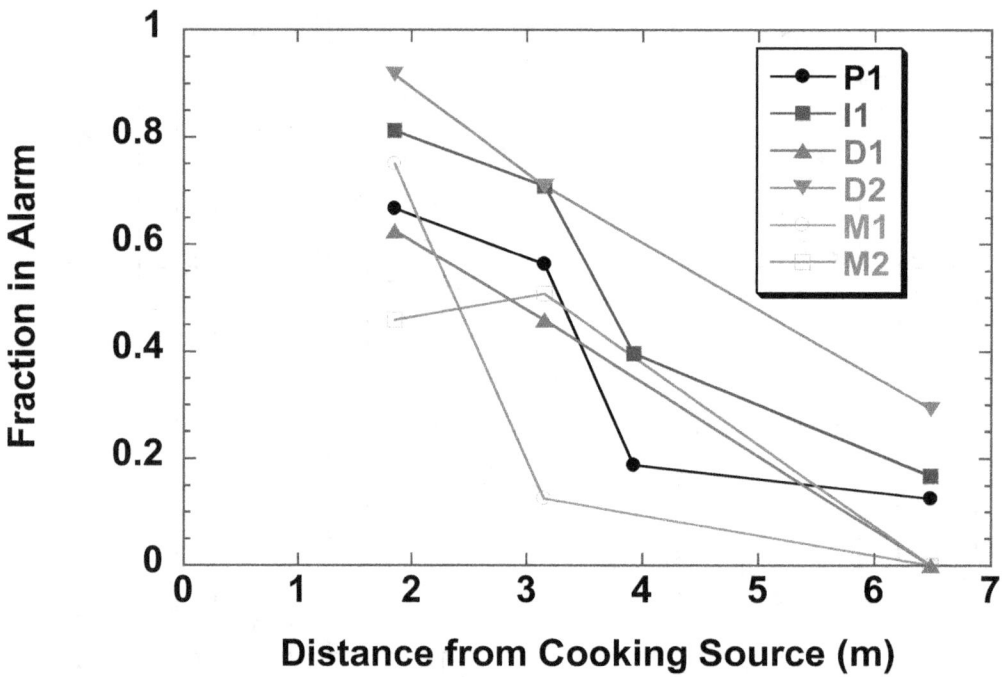

Figure 40. Fraction of smoke alarms that activated during range top cooking events.

Figure 41 shows the fraction of specific types of alarms activated during the aggregated electric oven events as a function of distance. M1 alarms activated during the oven cooking events whenever one was located in the kitchen, but none activated when located at the furthest two locations in the living room. The fraction of I1 and D2 alarms activated dropped when the location moved from the kitchen to just outside the kitchen, but remain the same at further distances. P1 alarms activated less than 20 % of the time they were present in the kitchen and recorded no alarm activations beyond 3.2 m from the range top.

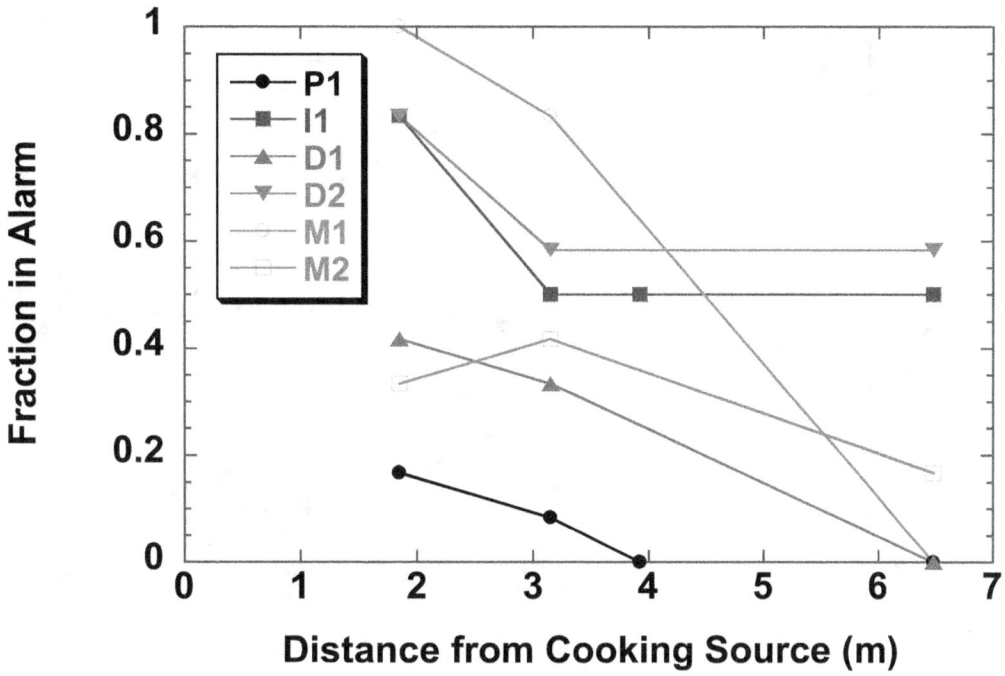

Figure 41. Fraction of smoke alarms that activated during oven cooking events.

Figure 42 shows the fraction of specific types of alarms activated during the aggregated toasting events as a function of distance. I1 and D2 activated over 75 % of the time they were present in the kitchen, while P1 activated approximately 25 % of the time. The fraction of alarm activations dropped as the distance from the cooking source increased for all alarm types.

While the frequency of nuisance alarms in actual usage depends on the frequency of the specific exposure events, it is still illustrative to present an averaged nuisance alarm response to cooking events. Instead of aggregating all cooking events, the fraction of alarms activated for the aggregated cooking activities, electric range, oven, and toasting were averaged so that each activity represented one third of all events. Figure 43 shows the alarm activation frequency for the three averaged cooking activities. Inside the kitchen, P1 has the lowest activation frequency, while I1 and D2 are both above 80 %. The alarm activation frequency drops as the distance from the range top increases. I1 and D2 exhibit higher activation frequencies than the other four alarms P1, D1, M1, and M2. Outside the kitchen those four alarm activation frequency values are similar at the different distances. These observations strengthen the case to keep smoke alarms outside of kitchens if possible, and if necessary, consider photoelectric type alarms. Outside the kitchen, the nuisance alarm performance of P1, D1, M1, and M2 is similar; they all appear significantly better than I1 and D2.

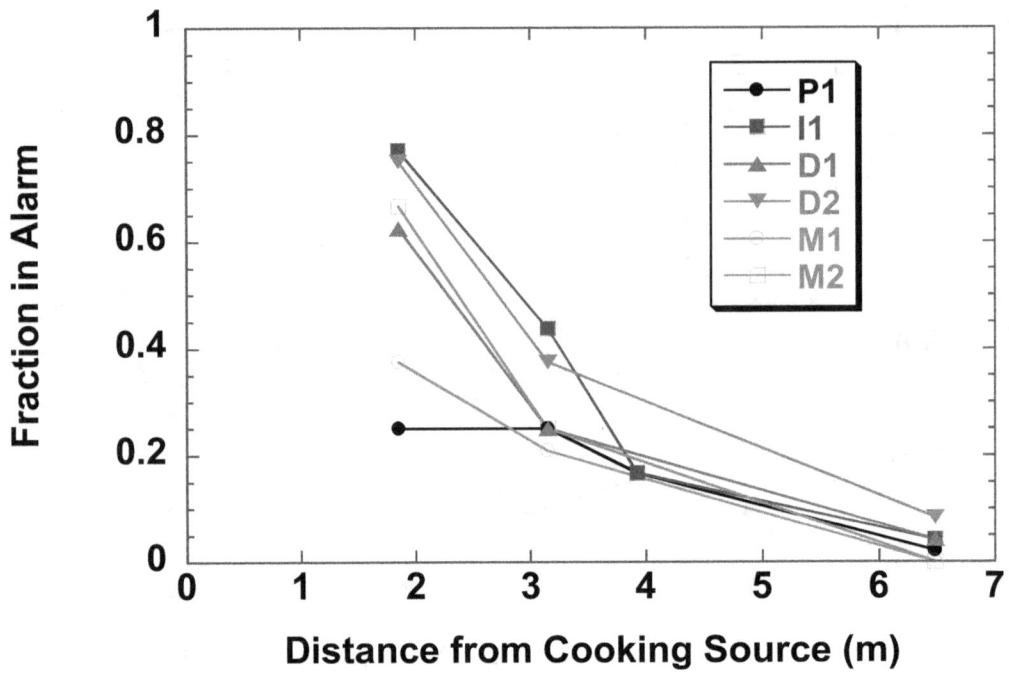

Figure 42. Fraction of smoke alarms that activated during toasting events.

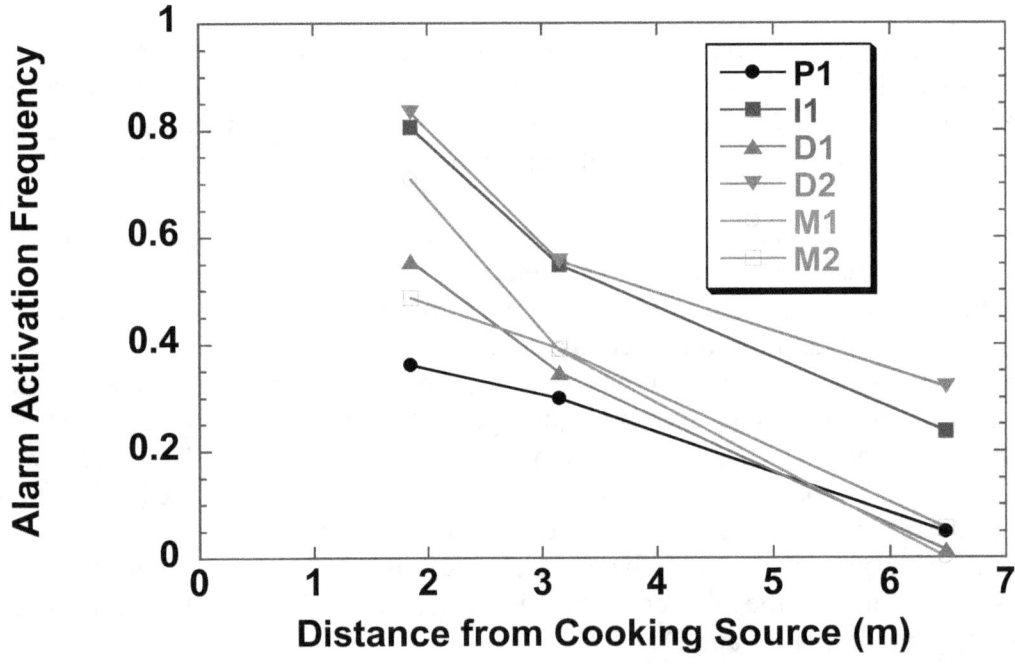

Figure 43. Alarm activation frequency for equal fractions of range top, oven and toasting activities.

3.4   Kitchen Fire Alarm Performance

A total of 10 fire tests were conducted, Table 44 identifies the configurations.

Test Name	Cabinet Construction	Ignition Scenario
A1_1	Oak/Pressboard	1
A1_2	Oak/Pressboard	1
A2_1	Oak/Pressboard	2
A2_2	Oak/Pressboard	2
B1_1	Laminated Pressboard	1
B1_2	Laminated Pressboard	1
B2_1	Laminated Pressboard	2
B2_2	Laminated Pressboard	2
A3_1	Oak/Pressboard, Sheet Metal Barrier	1
B3_1	Laminated Pressboard, Sheet Metal Barrier	1

Table 44. Configurations for kitchen fire tests.

For each test, up to 10 unused smoke alarms were installed on the ceiling of the burn prop building at various locations. The locations of the different types of smoke alarms were also varied from test-to-test. However, only photoelectric alarms were placed at Loc 1 and Loc 2 inside the kitchen to limit the potential for thermal damage of the ionization sensors in the other alarms.

The time to alarm was recorded for every smoke alarm installed in each test. The tenability conditions were assessed in the hallway, and the living room to determine if any given installed alarm provided sufficient time for egress. The tenability was assessed by considering the smoke optical density (OD) and the fractional effective dose (FED) of toxic gases or convected heat. The FED is a non-dimensional, time-integrated value of the exposure effects to toxic gases or convected and radiated heat that would be experienced by an occupant. The fractional effective dose (FED) calculation schemes are described in the standard ISO/FDIS 13751 [15]. A FED of 1.0 is associated with 50 % of exposed persons experiencing incapacitation and unable to effect escape. While the ISO standard does not include a FED incapacitation distribution, a FED value of 0.3 has been promoted as an exposure level that assures most occupants would not become incapacitated [15]. CO and CO_2 gas concentrations, and air temperature measurements at a height 1.5 m from the floor were used to calculate the toxic gas and convected heat FEDs.

Results for each test are presented in the tables (Table 45-54) and figures (Figure 44-71) that follow. Each table documents the time to alarm and alarm location for each alarm installed for a particular experiment. If the table entry is blank, no alarm was recorded during the test. If the table entry is gray, the particular alarm was not installed during that test. Next, a sequence of four images shows the fire progression at ignition, 120 s, 240 s, and 360 s after ignition, followed by an end-of-test picture of the cabinet for most tests. Lastly, the smoke optical density and the heat and toxic gases fractional effective dose (gases and temperature) for each measurement location are presented. The Y axis is scaled to the smoke optical density in m^{-1} and the toxic gas and heat FED which are dimensionless.

Distance from Stove, m (ft.)	P1 T_a (s)	P2 T_a (s)	I1 T_a (s)	I2 T_a (s)	D1 T_a (s)	D2 T_a (s)	M1 T_a (s)	M2 T_a (s)
1.82 (5.98)	241	153						
1.87 (6.12)								
2.96 (9.72)	243		162				184	
3.33 (10.93)	241		214		238			
4.50 (14.77)			209					236
5.39 (17.70)								
6.01 (19.71)								
6.94 (22.77)								

Table 45. Alarm times for experiment A1_1 (Oak/pressboard, scenario 1).

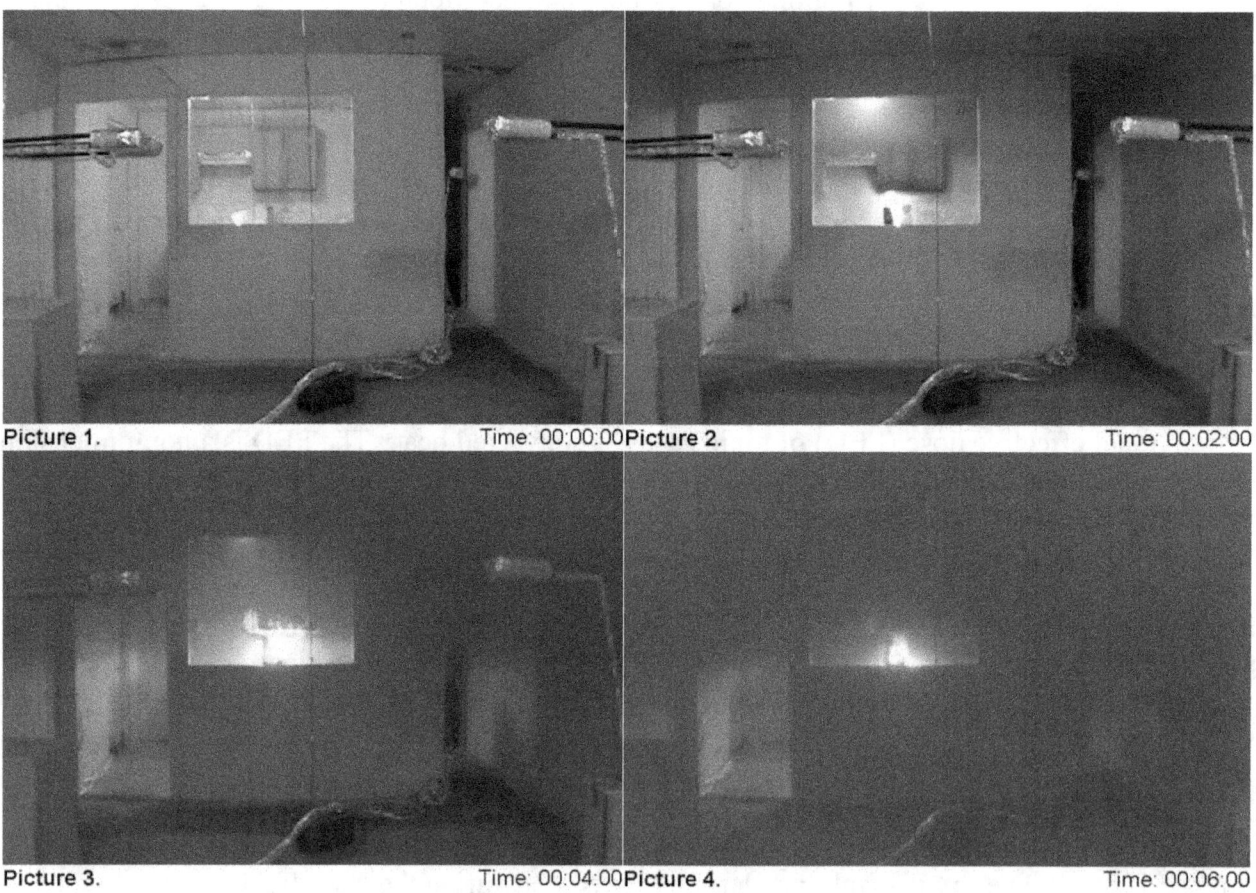

Figure 44. Photo sequence for experiment A1_1 (Oak/pressboard, scenario 1).

56

Figure 45. Post-fire photo of experiment A1_1 (Oak/pressboard, scenario 1).

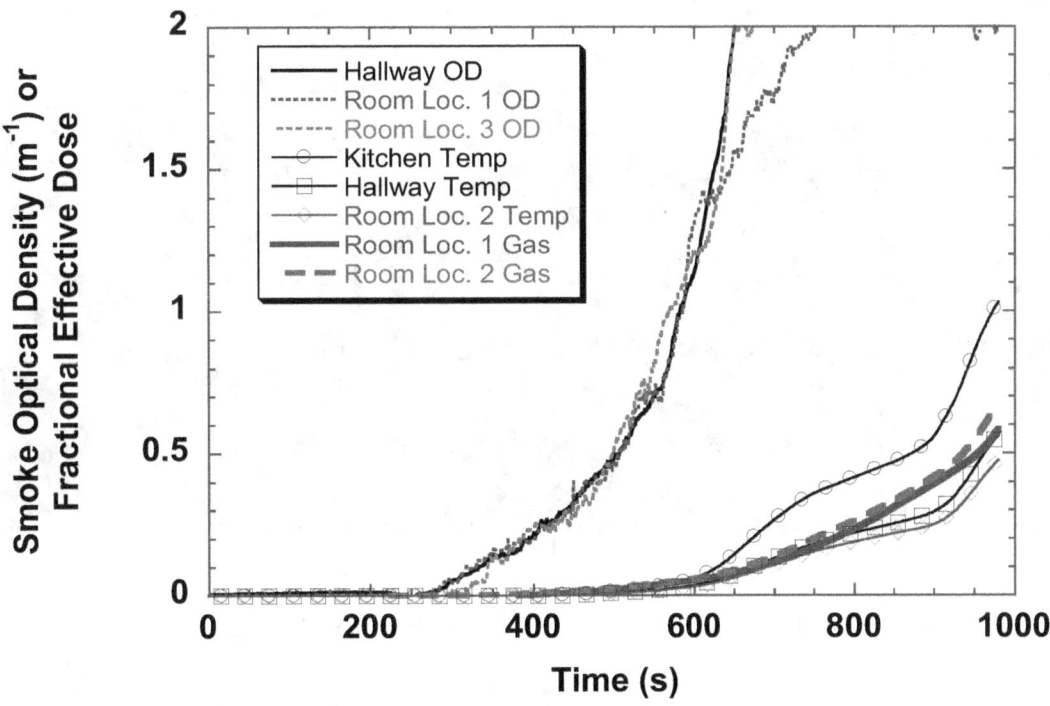

Figure 46. OD and FED values for experiment A1_1 (Oak/pressboard, scenario 1).

57

Distance from Stove, m (ft.)	P1 T_a (s)	P2 T_a (s)	I1 T_a (s)	I2 T_a (s)	D1 T_a (s)	D2 T_a (s)	M1 T_a (s)	M2 T_a (s)
1.82 (5.98)								
1.87 (6.12)	117	121						
2.96 (9.72)		133		127		129		
3.33 (10.93)								
4.50 (14.77)								
5.39 (17.70)								
6.01 (19.71)								
6.94 (22.77)	247		222				248	

Table 46. Alarm times for Experiment A1_2 (Oak/pressboard, scenario 1).

Picture 1.   Time: 00:00:00 Picture 2.   Time: 00:02:00

Picture 3.   Time: 00:04:00 Picture 4.   Time: 00:06:00

Figure 47. Photo sequence for experiment A1_2 (Oak/pressboard, scenario 1).

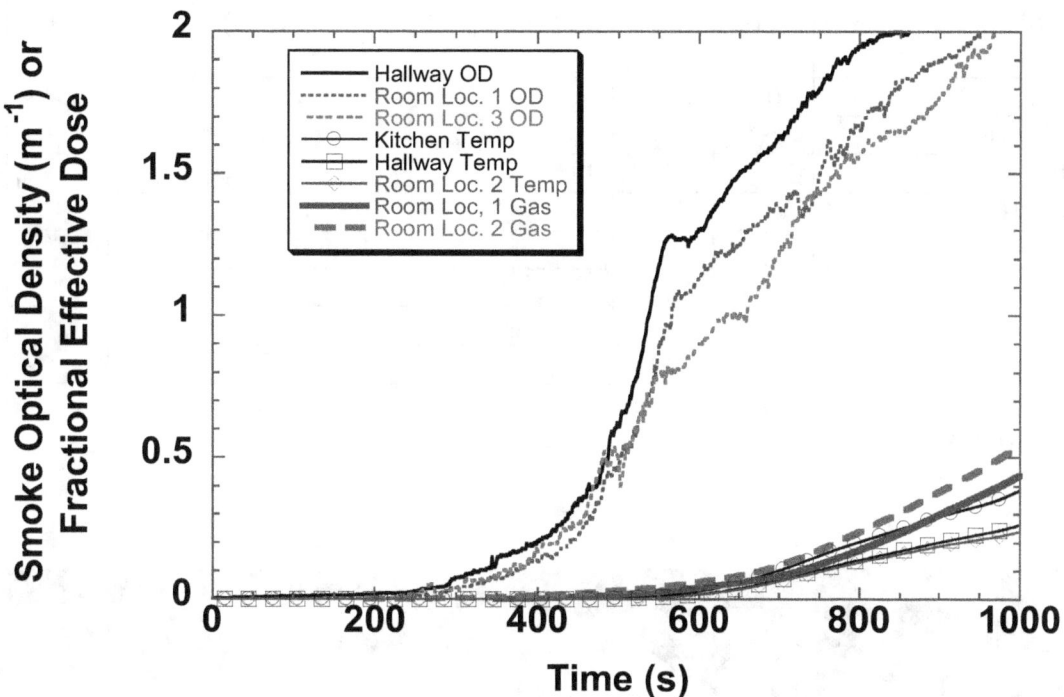

Figure 48. OD and FED values for experiment A1_2 (Oak/pressboard, scenario 1).

Distance from Stove, m (ft.)	P1 T_a (s)	P2 T_a (s)	I1 T_a (s)	I2 T_a (s)	D1 T_a (s)	D2 T_a (s)	M1 T_a (s)	M2 T_a (s)
1.82 (5.98)	104	98						
1.87 (6.12)								
2.96 (9.72)	115		115				122	
3.33 (10.93)								
4.50 (14.77)								
5.39 (17.70)								
6.01 (19.71)								
6.94 (22.77)		190		178		159		

Table 47. Alarm times for Experiment A2_1 (Oak/pressboard, scenario 2).

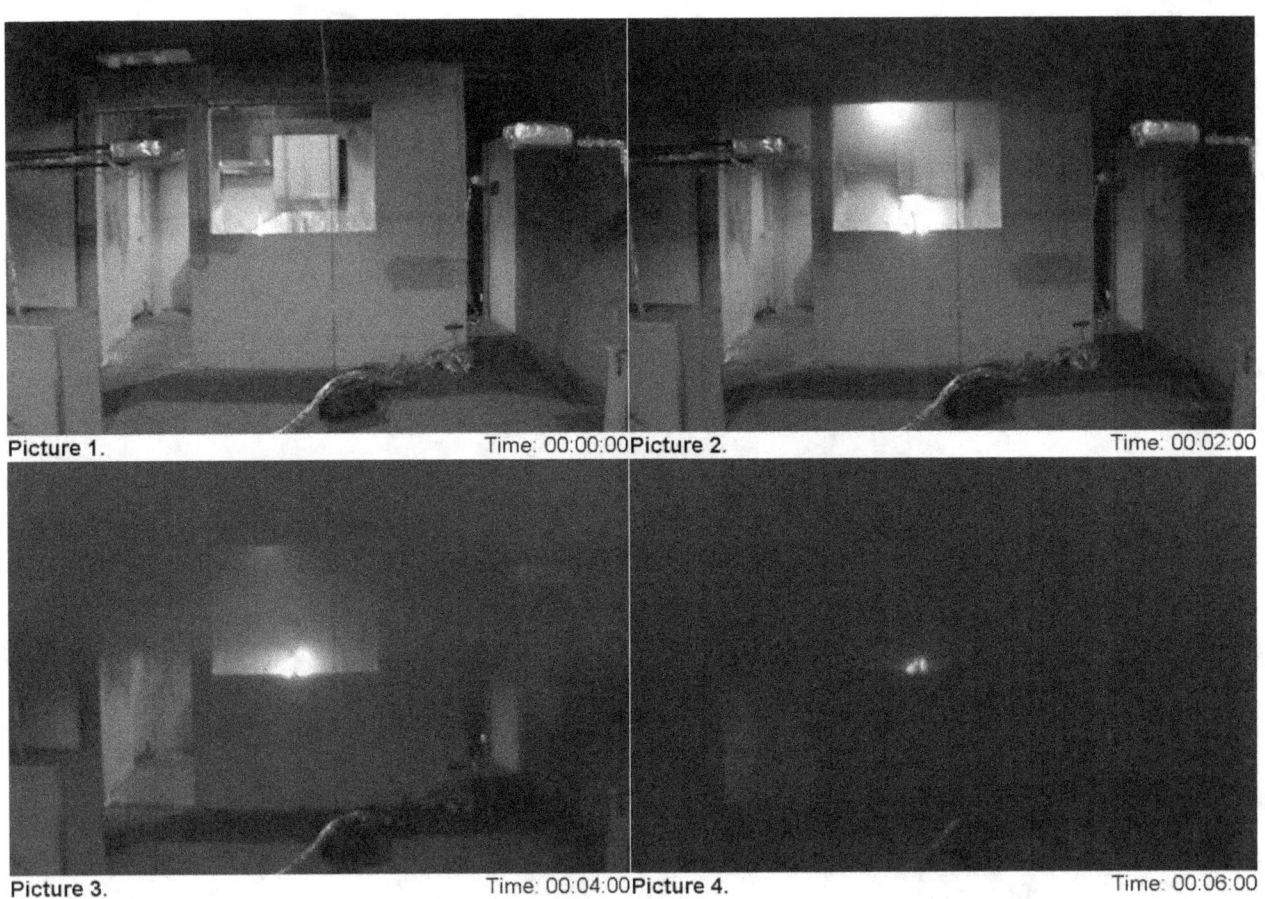

Picture 1.   Time: 00:00:00   Picture 2.   Time: 00:02:00

Picture 3.   Time: 00:04:00   Picture 4.   Time: 00:06:00

Figure 49. Photo sequence for experiment A2_1 (Oak/pressboard, scenario 2).

Figure 50. Post-fire photo of experiment A2_1 (Oak/pressboard, scenario 2).

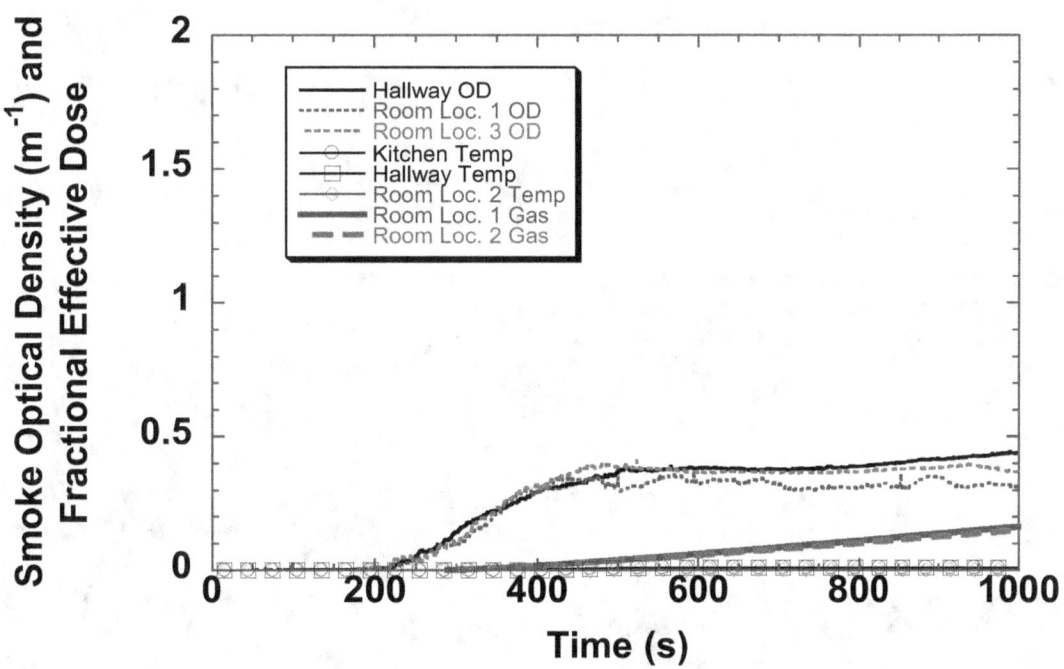

Figure 51. OD and FED values for experiment A2_1 (Oak/pressboard, scenario 2).

Distance from Stove, m (ft.)	P1 T_a (s)	P2 T_a (s)	I1 T_a (s)	I2 T_a (s)	D1 T_a (s)	D2 T_a (s)	M1 T_a (s)	M2 T_a (s)
1.82 (5.98)	125	119						
1.87 (6.12)								
2.96 (9.72)								
3.33 (10.93)	205		154		181			
4.50 (14.77)			159					169
5.39 (17.70)								
6.01 (19.71)								
6.94 (22.77)								

Table 48. Alarm times for experiment A2_2 (Oak/pressboard, scenario 2).

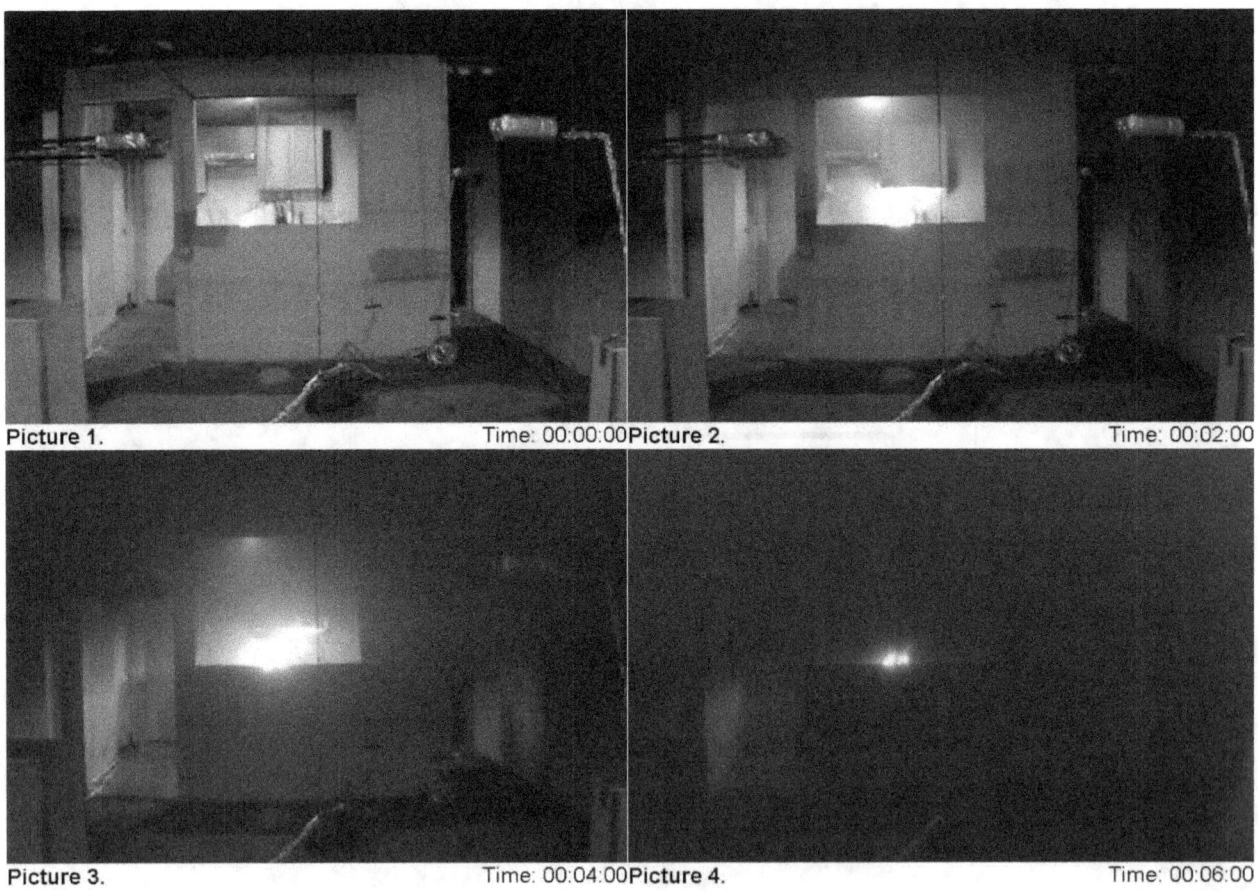

Picture 1.   Time: 00:00:00 Picture 2.   Time: 00:02:00

Picture 3.   Time: 00:04:00 Picture 4.   Time: 00:06:00

Figure 52. Photo sequence for experiment A2_2 (Oak/pressboard, scenario 2).

Figure 53. Post-fire photo of experiment A2_2 (Oak/pressboard, scenario 2).

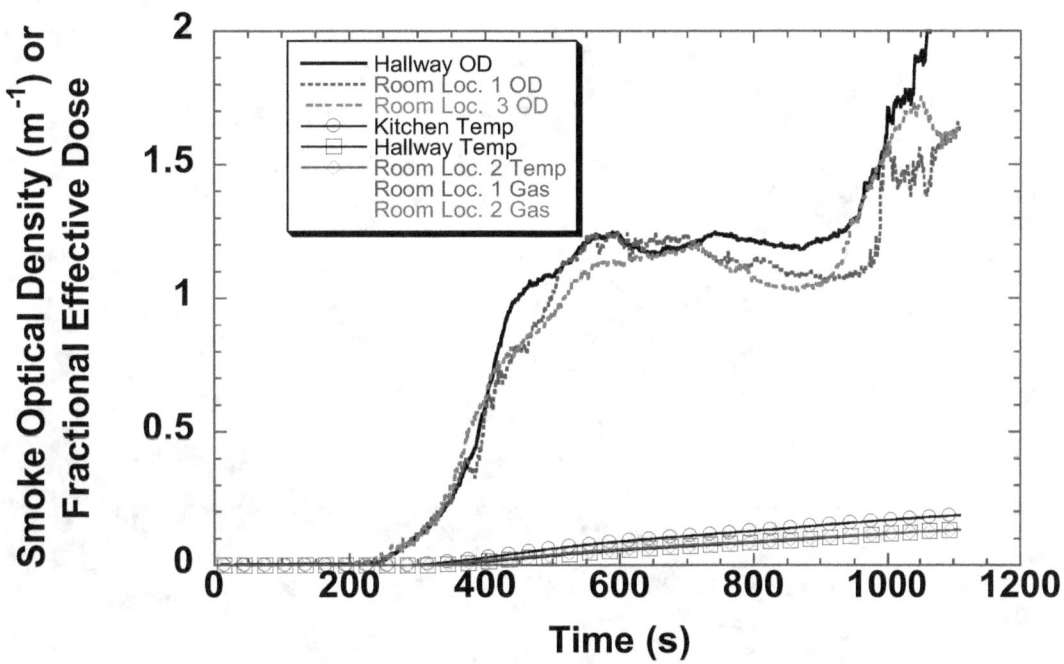

Figure 54. OD and FED values for experiment A2_2 (Oak/pressboard, scenario 2).

Distance from Stove, m (ft.)	P1 T_a (s)	P2 T_a (s)	I1 T_a (s)	I2 T_a (s)	D1 T_a (s)	D2 T_a (s)	M1 T_a (s)	M2 T_a (s)
1.82 (5.98)	173	167						
1.87 (6.12)								
2.96 (9.72)			123			128	140	
3.33 (10.93)	182		142		180			
4.50 (14.77)			145					168
5.39 (17.70)								
6.01 (19.71)								
6.94 (22.77)								

Table 49. Alarm times for Experiment B1_1 (Laminated pressboard, scenario 1).

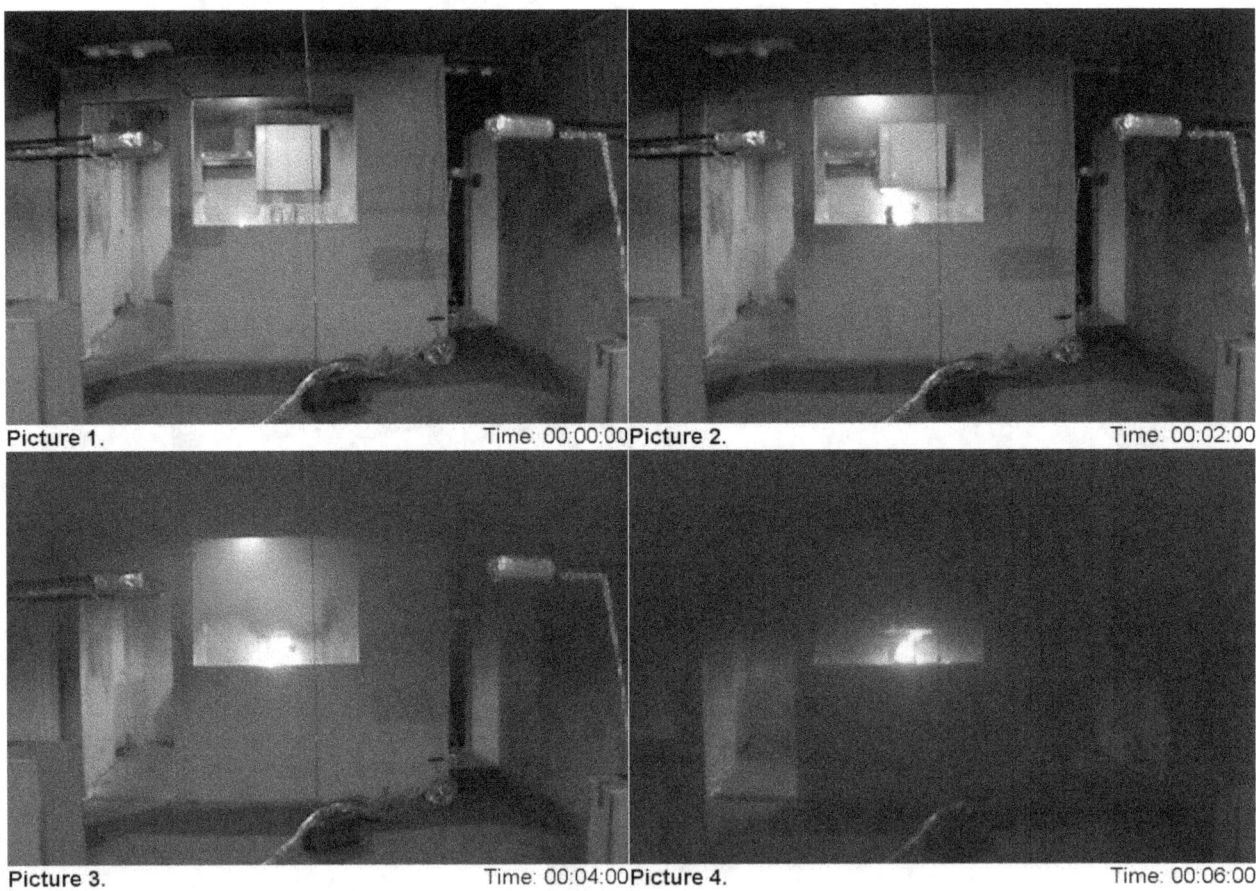

Picture 1.   Time: 00:00:00 Picture 2.   Time: 00:02:00

Picture 3.   Time: 00:04:00 Picture 4.   Time: 00:06:00

Figure 55. Photo sequence for experiment B1_1 (Laminated pressboard, scenario 1).

Figure 56. Post-fire photo of experiment B1_1 (Laminated pressboard, scenario 1).

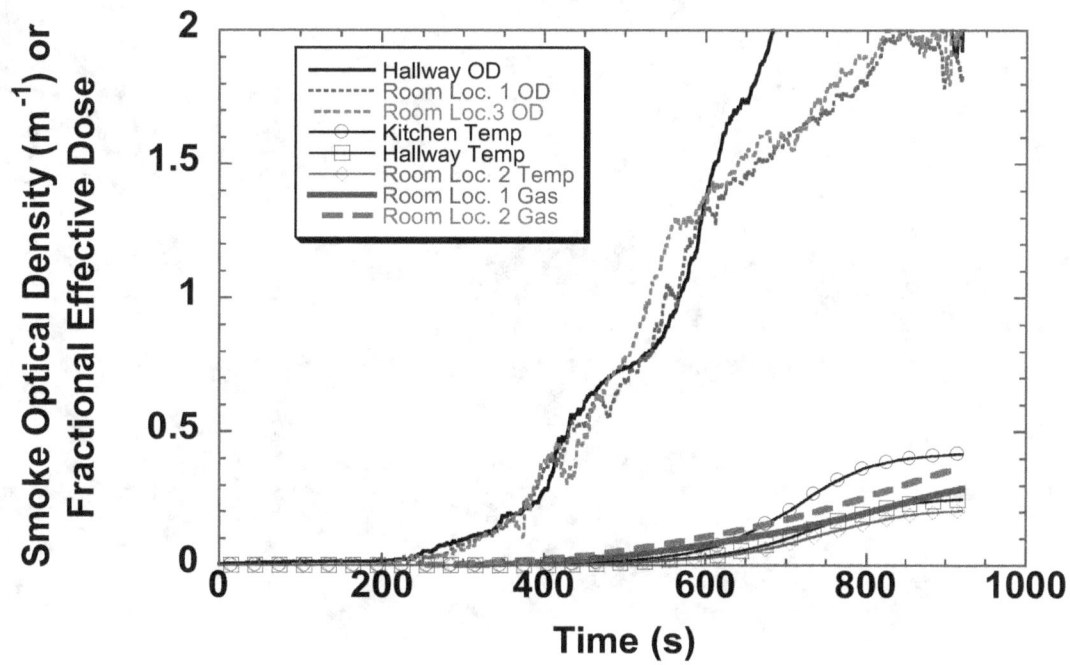

Figure 57. OD and FED values for experiment B1_1 (Laminated pressboard, scenario 1).

Distance from Stove, m (ft.)	P1 T_a (s)	P2 T_a (s)	I1 T_a (s)	I2 T_a (s)	D1 T_a (s)	D2 T_a (s)	M1 T_a (s)	M2 T_a (s)
1.82 (5.98)								
1.87 (6.12)	113	140						
2.96 (9.72)	121		117				128	
3.33 (10.93)								
4.50 (14.77)								
5.39 (17.70)								
6.01 (19.71)								
6.94 (22.77)	193			163		156		

Table 50. Alarm times for Experiment B1_2 (Laminated pressboard, scenario 1).

Figure 58. Photo sequence for experiment B1_2 (Laminated pressboard, scenario 1).

Figure 59. Post-fire photo of experiment B1_2 (Laminated pressboard, scenario 1).

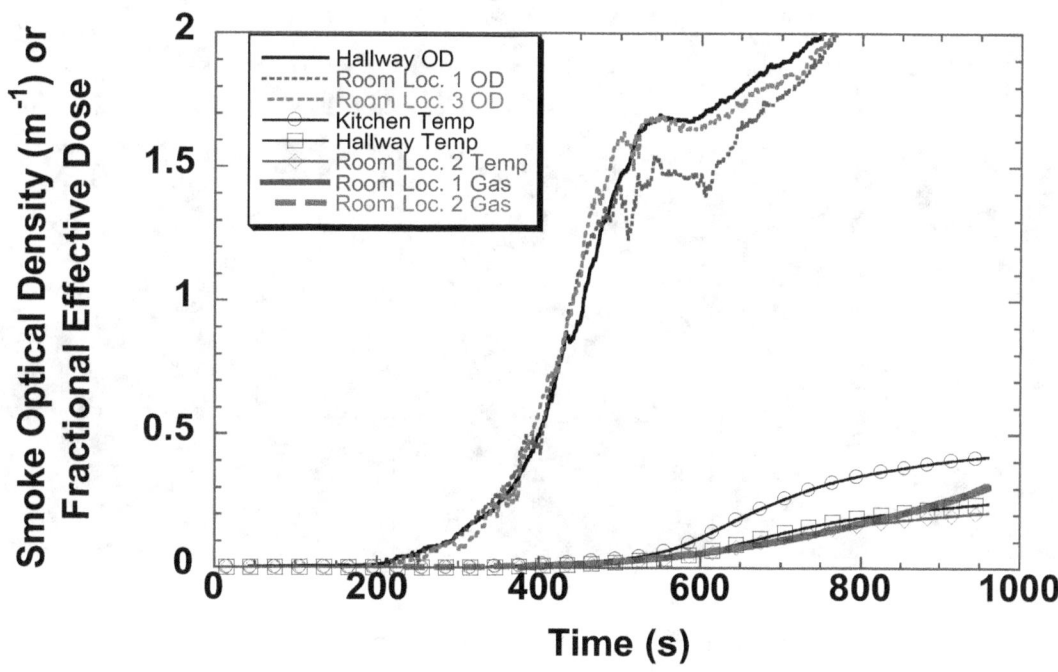

Figure 60. OD and FED values for experiment B1_2 (Laminated pressboard, scenario 1).

Distance from Stove, m (ft.)	P1 T_a (s)	P2 T_a (s)	I1 T_a (s)	I2 T_a (s)	D1 T_a (s)	D2 T_a (s)	M1 T_a (s)	M2 T_a (s)
1.82 (5.98)	126	127						
1.87 (6.12)								
2.96 (9.72)		140		137		134		
3.33 (10.93)			150					170
4.50 (14.77)								
5.39 (17.70)								
6.01 (19.71)								
6.94 (22.77)	232		193				211	

Table 51. Alarm times for Experiment B2_1 (Laminated pressboard, scenario 2).

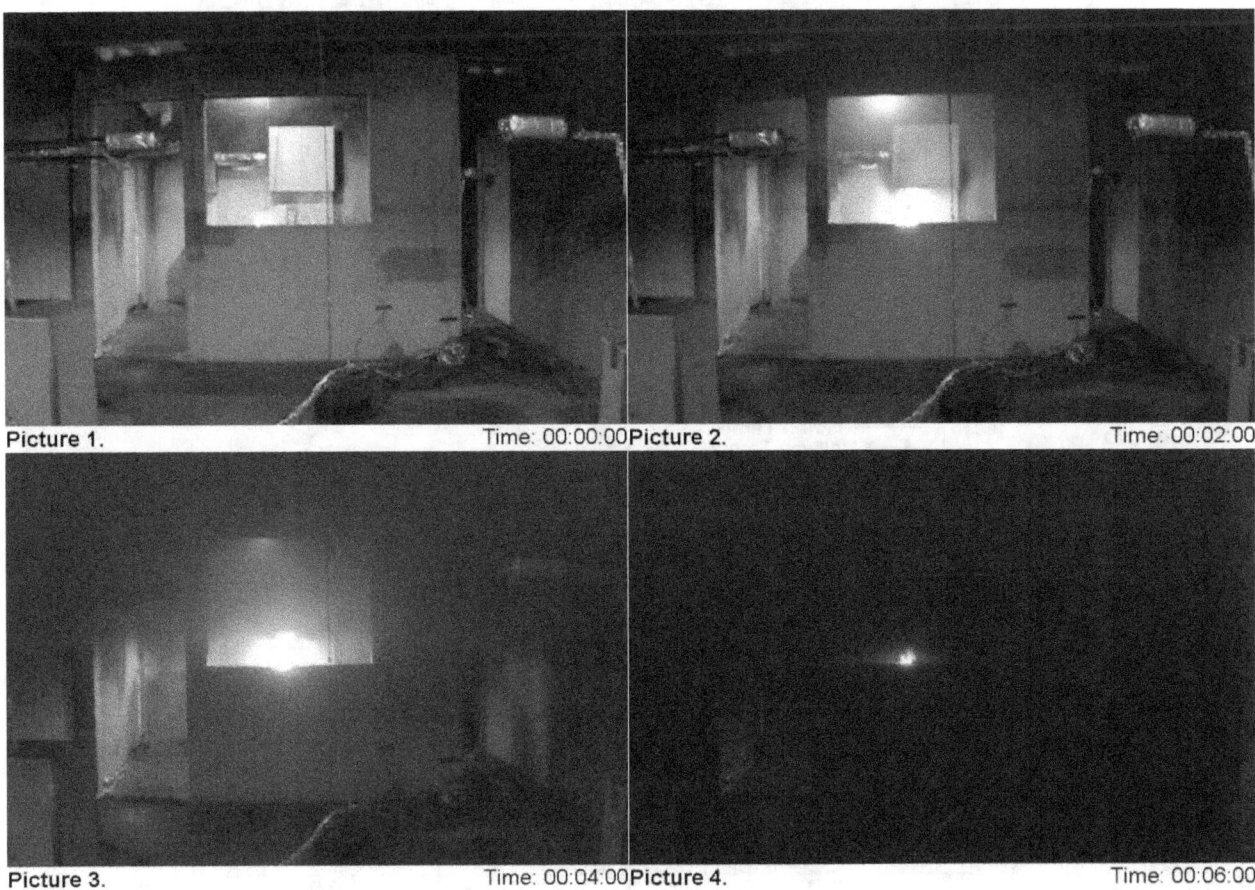

Picture 1.   Time: 00:00:00 Picture 2.   Time: 00:02:00

Picture 3.   Time: 00:04:00 Picture 4.   Time: 00:06:00

Figure 61. Photo sequence for experiment B2_1 (Laminated pressboard, scenario 2).

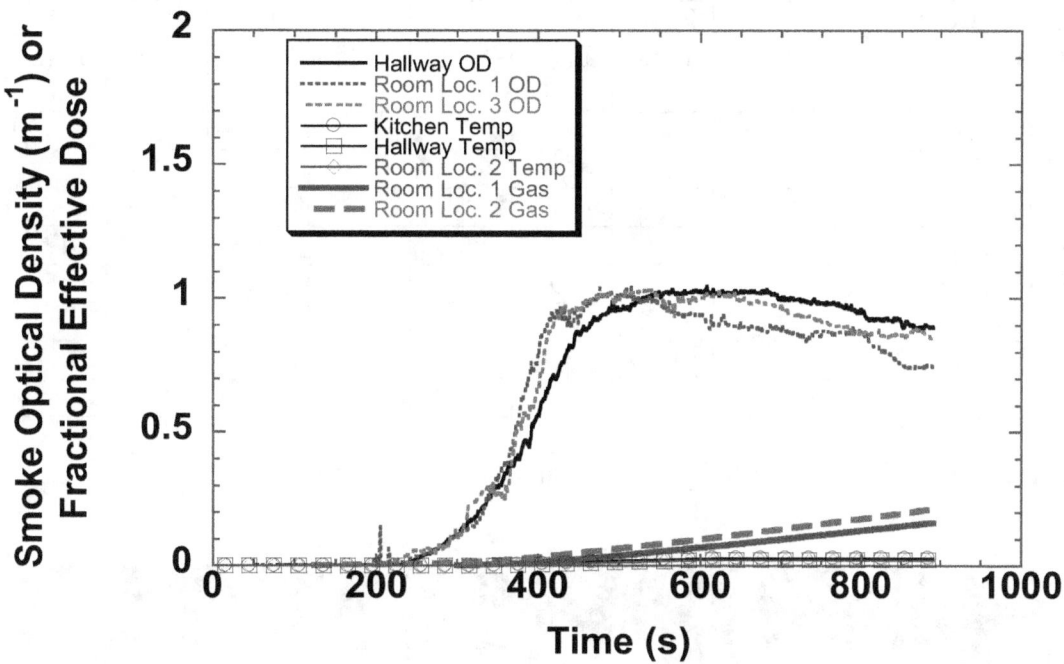

Figure 62. OD and FED values for experiment B2_1 (Laminated pressboard, scenario 2).

Distance from Stove, m (ft.)	P1 T_a (s)	P2 T_a (s)	I1 T_a (s)	I2 T_a (s)	D1 T_a (s)	D2 T_a (s)	M1 T_a (s)	M2 T_a (s)
1.82 (5.98)								
1.87 (6.12)	147	161						
2.96 (9.72)			117					
3.33 (10.93)								149
4.50 (14.77)			139					145
5.39 (17.70)								
6.01 (19.71)								
6.94 (22.77)		221						

Table 52. Alarm times for Experiment B2_2 (Laminated pressboard, scenario 2).

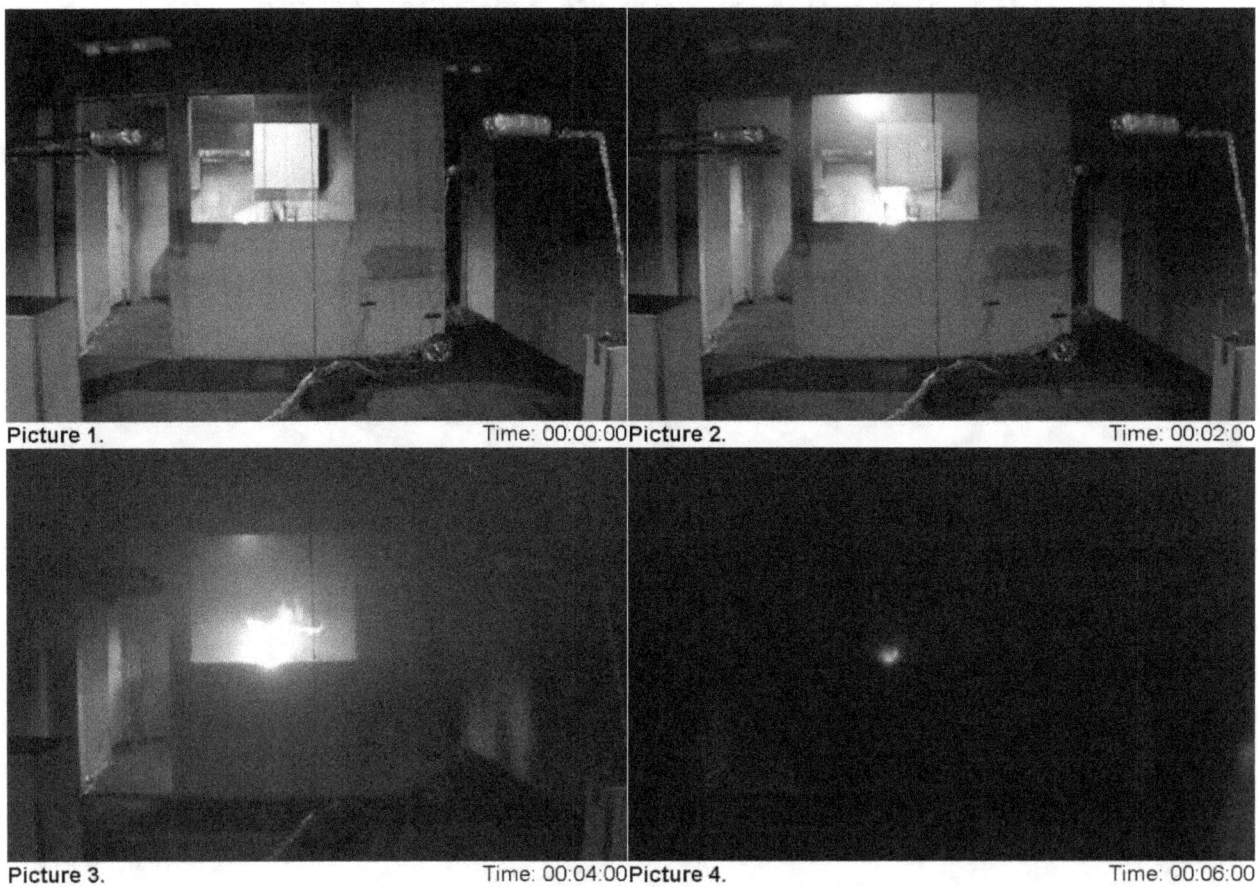

Figure 63. Photo sequence for experiment B2_2 (Laminated pressboard, scenario 2).

Figure 64. Post-fire photo of experiment B2_2 (Laminated pressboard, scenario 2).

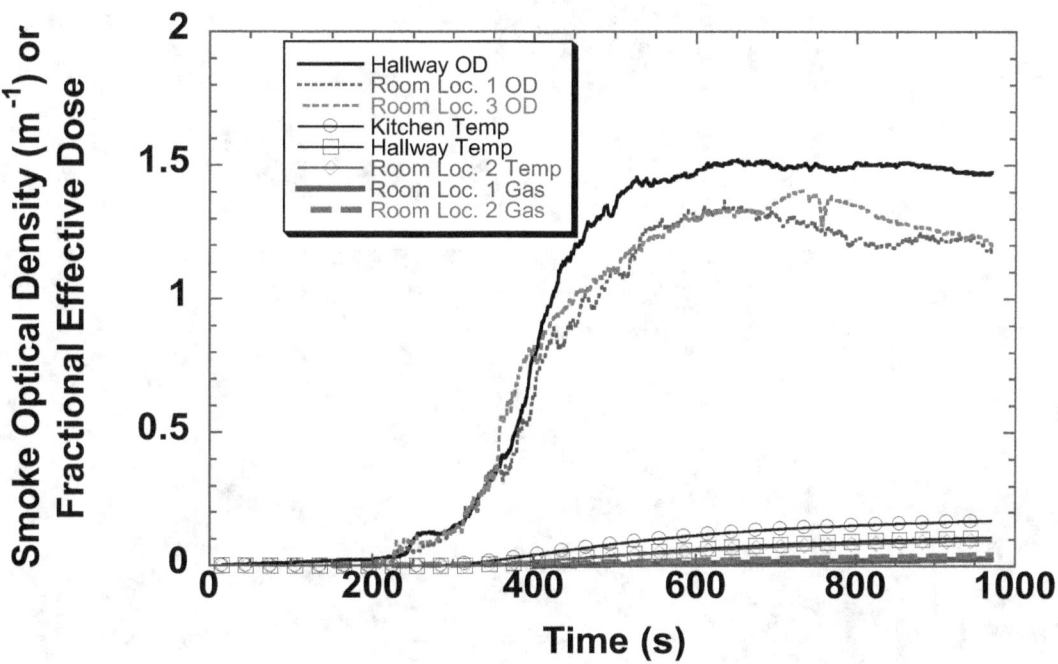

Figure 65. OD and FED values for experiment B2_2 (Laminated pressboard, scenario 2).

Distance from Stove, m (ft.)	P1 T_a (s)	P2 T_a (s)	I1 T_a (s)	I2 T_a (s)	D1 T_a (s)	D2 T_a (s)	M1 T_a (s)	M2 T_a (s)
1.82 (5.98)								
1.87 (6.12)	121	130						
2.96 (9.72)	127		125				133	
3.33 (10.93)								
4.50 (14.77)								
5.39 (17.70)								
6.01 (19.71)								
6.94 (22.77)	219		203		213			

Table 53. Alarm times for Experiment A3_1 (Oak/pressboard with sheet metal barrier, scenario 1).

Figure 66. Photo sequence for experiment A3_1 (Oak/pressboard with sheet metal barrier, scenario 1).

Figure 67. Post-fire photo of experiment A3_1 (Oak/pressboard with sheet metal barrier, scenario 1).

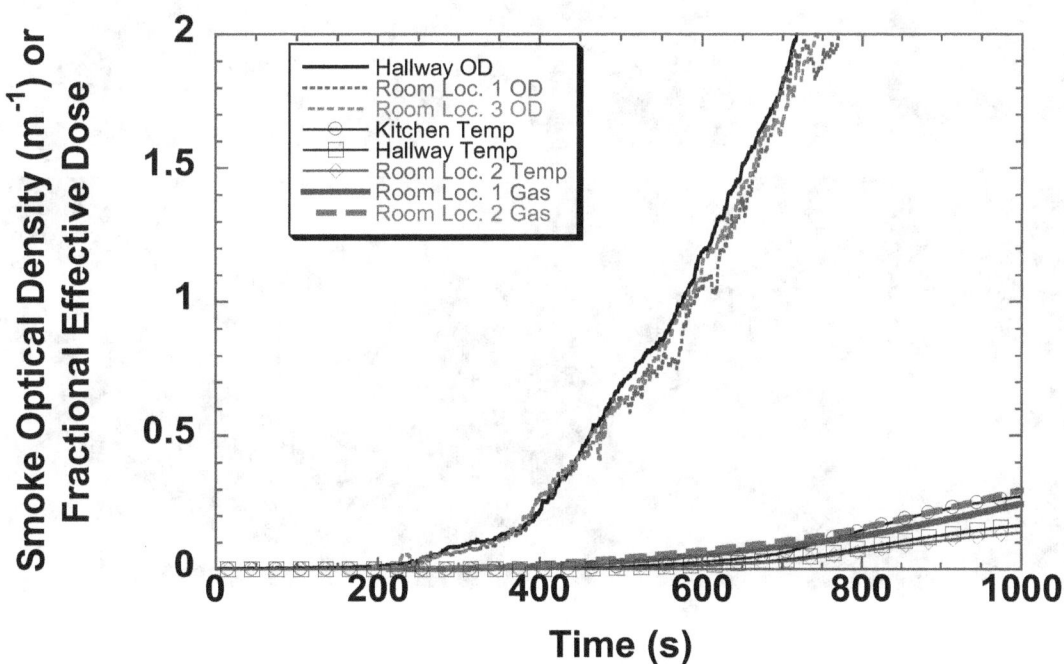

Figure 68. OD and FED values for experiment A3_1 (Oak/pressboard with sheet metal barrier, scenario 1).

Distance from Stove, m (ft.)	P1 T_a (s)	P2 T_a (s)	I1 T_a (s)	I2 T_a (s)	D1 T_a (s)	D2 T_a (s)	M1 T_a (s)	M2 T_a (s)
1.82 (5.98)		125						
1.87 (6.12)								
2.96 (9.72)			133		151			
3.33 (10.93)								
4.50 (14.77)			142					141
5.39 (17.70)								
6.01 (19.71)								
6.94 (22.77)			197				217	

Table 54. Alarm times for Experiment B3_1 (Laminated pressboard with sheet metal barrier, scenario 1).

Figure 69. Photo sequence for experiment B3_1 (Oak/pressboard with sheet metal barrier, scenario 1).

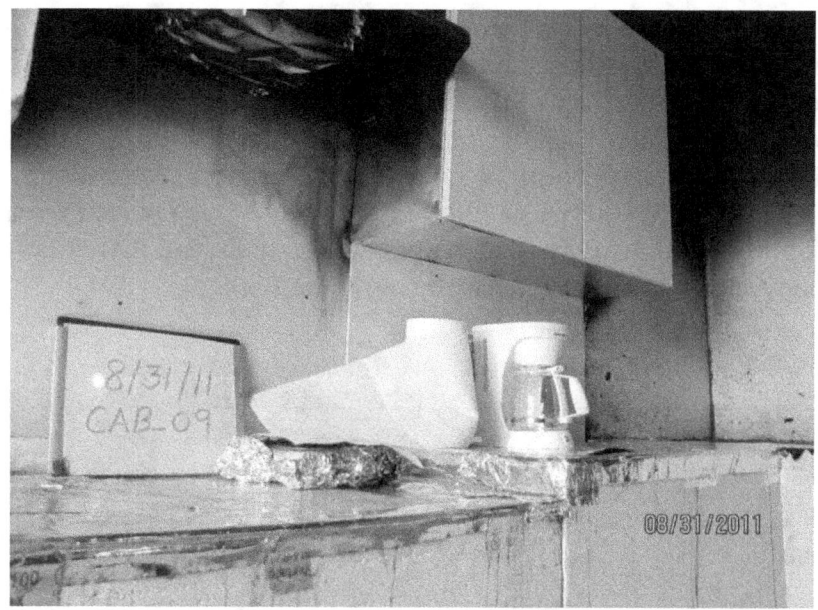

Figure 70. Pre-fire photo of experiment B3_1 (Oak/pressboard with sheet metal barrier, scenario 1).

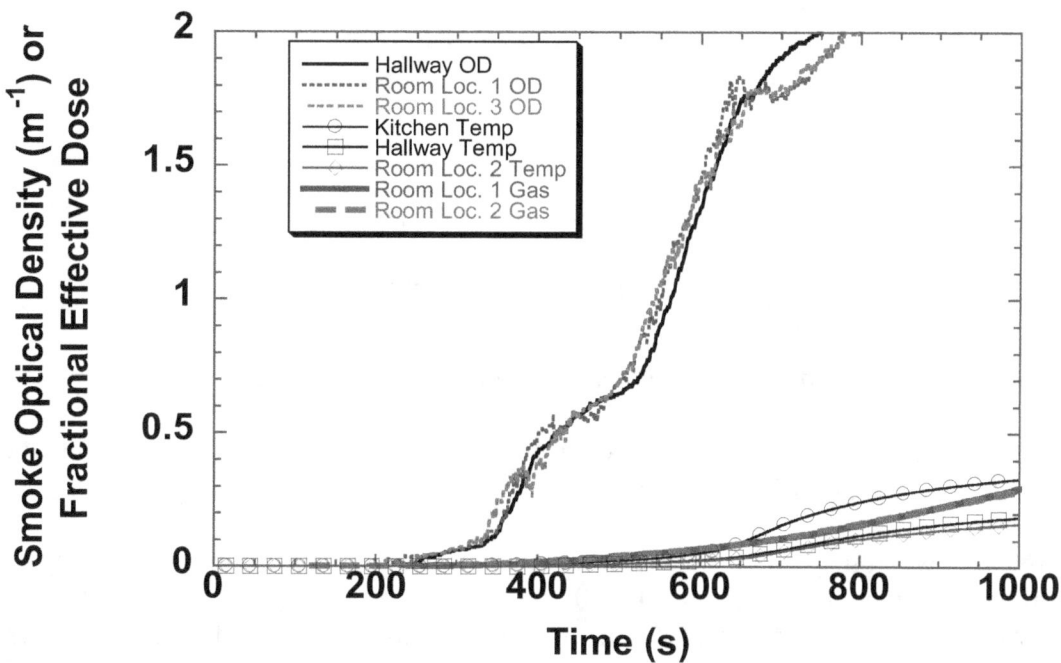

Figure 71. OD and FED values for experiment B3_1 (Oak/pressboard with sheet metal barrier, scenario 1).

Every installed smoke alarm activated during the 10 experiments. Typically, a photoelectric alarm in the kitchen was the first to activate, and all smoke alarms activated within about 100 s of the first alarm activation. Since the early stage of fire growth was similar for the two ignition scenarios, similarity in alarm time range for all 10 experiments is not surprising.

75

At locations outside the kitchen where I1 was present, it alarmed first 16 out of 19 times. In one case M2 alarmed 1 s before I1 and in another, P1 alarmed at the same time as I1. Comparing the difference between the alarm times of P1, D1, M1, and M2 versus the alarm time of a collocated I1 alarm a relative sensitivity ranking is obtained. Figure 72 shows the average difference for the four alarms and I1. There were 5 to 10 observations for each alarm, and the error bars represent ± one standard deviation. The range in alarm times for I1 was 115 s to 214 s. While the averages suggest an increasing sensitivity trend of P1-D1-M1-M2 to these kitchen fires, the magnitude is small, and there may not be a statistically significant difference between some alarm sensitivities to these fires.

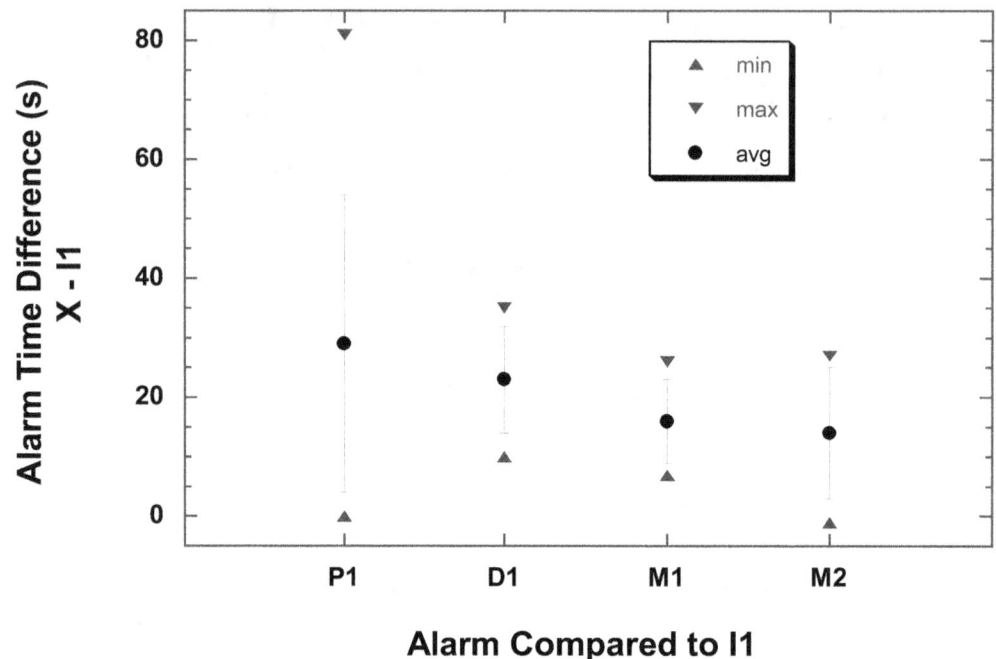

Figure 72. Average alarm time difference between Alarm type and collocated I1 alarm.

In order to compare the smoke alarm performance, an FED limiting value of 0.3 was chosen and two limiting smoke optical densities 0.25 m^{-1} and 0.50 m^{-1} were considered.

While there were cases when the FED for toxic gases or heat reached a limiting value of 0.3, smoke optical density always reached values greater than 0.25 m^{-1} well before any FED limit. The smoke optical density reached values greater than 0.50 m^{-1} well before any FED limit for all experiments except A2_1 (Oak/pressboard, ignition scenario 2) which never reached an optical density limit of 0.50 m^{-1} or a FED of 0.3. Table 55 shows the time to the first and last alarm activation for each experiment, and the time to reach the smoke optical density limits of 0.25 m^{-1} or 0.50 m^{-1} for the three extinction meter locations. The last alarm activation time was always before the time needed to reach the 0.25 m^{-1} optical density limit; thus, even with the slowest alarm activation, there was time to egress before significant smoke obscuration. Thus the available safe egress time (ASET) (defined as time to a FED or Smoke limit (whichever is reached first) minus the time to alarm) was positive. Any ASET for a particular smoke alarm activation time and time to reach a chosen smoke optical density limit can be computed using the tabulated values.

Test	First Alarm (s)	Last Alarm (s)	Time to Smoke OD Hallway		Time to Smoke OD Room Loc. 1		Time to Smoke OD Room Loc. 3	
			0.25 m^{-1}	0.50 m^{-1}	0.25 m^{-1}	0.50 m^{-1}	0.25 m^{-1}	0.50m^{-1}
A1_1	153	243	408	506	419	504	402	498
A1_2	117	248	423	486	455	502	433	480
A2_1	98	190	368	-	360	-	358	-
A2_2	119	205	348	390	349	396	340	375
B1_1	123	182	395	429	380	430	384	453
B1_2	113	193	351	399	342	390	353	391
B2_1	127	232	339	392	335	371	327	373
B2_2	117	221	330	380	332	376	324	357
A3_1	121	219	403	461	395	471	396	462
B3_1	125	217	371	424	369	400	350	424

Table 55. Tabulated first and last alarm activation time and time to reach threshold smoke optical densities.

Figure 73 shows ASET comparisons for four different cases, the difference between the time to reach an optical density limit of either 0.25 m^{-1} or 0.50 m^{-1} first in either the room or hallway locations and either the first or last alarm activation time. Thus, the shortest ASET was computed by using the last alarm activation time and the time to reach an optical density limit of 0.25 m^{-1} and the longest ASET was computed by using the first alarm activation time and the time to reach the optical density limit of 0.50 m^{-1}.

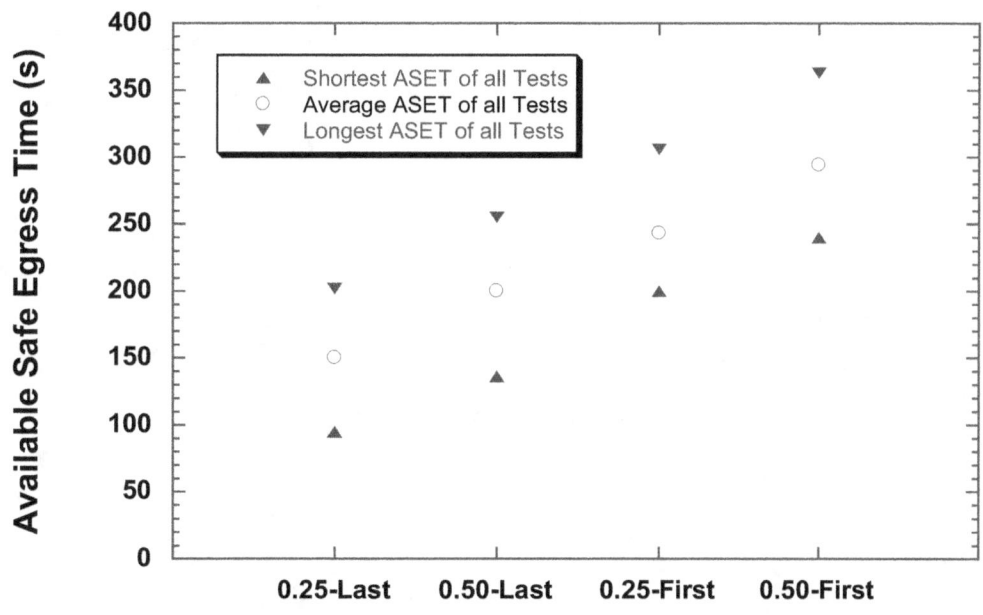

Figure 73. ASET computed using first or last alarm activation and time to reach 0.25 m^{-1} or 0.50 m^{-1}

optical density limit.

The difference between the average ASET using alarm times from the first and last smoke alarms to activate was approximately 100 s for both optical density limits. Comparing the results for the shortest ASET computed for each experiment (that is, the last smoke alarm to activate and an optical density limit of 0.25 m^{-1}), it varied from 95 s to 198 s with an average value of 150 s for all ten experiments. Only three computed ASET values were less than 120 s, two P1 alarms and one M1 alarm located at the furthest distance from the kitchen. While all of the smoke alarm provided ASET values greater than 94 s or 135 s given an optical density limit of 0.25 m^{-1} or 0.50 m^{-1}, the results do reflect the need to place smoke alarms in central locations in order to adequately detect all fires in the protected space.

4   Conclusions

The data collected provides insight into the fire growth and hazard development of kitchen fires, susceptibility of smoke alarms to cooking nuisance sources, and smoke alarm performance in kitchen fires. It is important to note that the overall performance of smoke alarms in residential settings is not limited to kitchen fire detection, but includes a range of fire scenarios. There is a fairly extensive body of research documenting alarm performance on a wide variety of fire scenarios that must be considered in order to assess overall alarm performance. Specifically, the performance of M1 and M2 (the multi-sensor, intelligent alarms) to a range of fire scenarios including smoldering fires, has not been documented.

Several conclusions can be drawn from the experimental results. From the kitchen fire scenario heat release rate measurements, the following conclusions are drawn.

1. Combustible materials typically found on a countertop can spread flames to overhead cabinets.
2. A single cabinet can produce a peak heat release rate nearly sufficient to flashover a small room.
3. A protective barrier on the bottom and side facing the range may limit the spread of flames to the cabinet and tends to reduce the heat release rate.

From the kitchen nuisance alarm tests the following conclusions are drawn.

1. For the conditions studied here, the propensity to nuisance alarm decreased as the distance from the cooking source increased.
2. Alarms (ionization alarm I1 and dual sensor alarm D2) that rely on sensitive ionization chambers experienced significantly more nuisance alarm activations for cooking activities and locations tested in this study.
3. All alarms except I1 and D2 experienced about the same nuisance alarm frequency for the locations outside the kitchen for the cooking scenarios tested.

From the kitchen fire tests the following conclusions are drawn.

1. Smoke optical densities always reached values greater than 0.25 m^{-1} well before the fractional equivalent dosage (FED) limit obtained a value of 0.3 in all the kitchen fires studied here.
2. All smoke alarms responded before hazardous conditions developed for the scenarios tested here; the I1 alarm tended to respond first at a given location.
3. Some smoke alarms placed at the furthest location from the kitchen (6 m) did not provide adequate available safe egress times (ASET) for the fire scenarios tested here, where adequate ASET times were defined in terms of an optical density limit of 0.25 m^{-1}.

The location requirements specified in the NFPA 72 Code appear to reduce potential nuisance alarm problems, but do not guarantee that nuisance alarms would be not problematic in all situations. The 2013 edition of the National Fire Alarm and Signaling Code has included language requiring smoke alarms and smoke detectors used in household fire alarm systems installed near cooking appliances be listed for resistance to common nuisance sources from cooking by 2016. Research at NIST and elsewhere is being conducted to develop specific tests (including cooking sources) for nuisance resistant alarms tailored to remove the most egregious alarms with the goal of improving smoke alarm nuisance resistance performance.

5   References

1. Karter, M.J., "Fire Loss in the United States 2008," National Fire Protection Association, Quincy, MA., September, 2009.
2. Ahrens, M., "Home Smoke Alarms – The Data as Context for Decision," *Fire Technology*, January 2008, DOI: 10.1007/s10694-008-0045-9.
3. Ahrens, M., "Smoke Alarms in US Home Fires," National Fire Protection Association, Quincy, MA., September, 2011.
4. Rowland D.,et al. (2002) Prevalence of working smoke alarms in local authority inner city housing: randomized controlled trial. *BMJ, 325,* 998-1001.
5. Mueller, B.A., Sidman, E.A., Alter, H., Perkins, R., and Grossman, D.C., (2008) Randomized Controlled Trial of Ionization and Photoelectric Smoke Alarm Functionality, *Inj Prev* 2008;14:80–86. doi:10.1136/ip.2007.016725
6. Yang, J., Jones, M.P., Cheng, G., Ramirez, M., Taylor, C., Peek-Asa, C., (2011) Do Nuisance Alarms Decrease Functionality of Smoke alarms Near the Kitchen? Findings from a Randomized Controlled Study, *Inj Prev* doi:10.1136/ip.2010.027805
7. Ahrens, M., "Home Fires Involving Cooking Equipment," National Fire Protection Association, Quincy, MA., November, 2011.
8. R. W. Bukowski, R. D. Peacock, J. D. Averill, T. G. Cleary, N. P. Bryner, W. D. Walton, P. A. Reneke, and E. D. Kuligowski, NIST Technical Note 1455, *Performance of Home Smoke Alarms: Analysis of the Response of Several Available Technologies in Residential Fire Settings,* Washington, DC: U.S. Department of Commerce, National Institute of Standards and Technology, 2008 revision.
9. Mealy, C., Wolfe, A., and Gottuk, D., "Smoke Alarm Response and Tenability," AUBE 09, 14th International Conf. on Automatic Fire Detection, Duisburg Germany, Sept 8-10, 2009.
10. Lee, A., and Pineda D., "Smoke Alarms – Pilot Study of Nuisance Alarms Associated with Cooking," US Consumer Products Safety Commission, Bethesda, MD, March 2010.
11. NFPA 72, National Fire Alarm and Signaling Code, 2010 Edition, NFPA, Quincy MA.
12. UL 217 Single and Multiple Station Smoke Alarms, Underwriters Laboratories, Northbrook IL.
13. Bryant, R., Ohlemiller, T., Johnsson, E., Hamins, A., Grove, B., Guthrie, W.F., Maringhides, A.and Mulholland, G., The NIST 3 Megawatt Quantitative Heat Release Rate Facility, *NIST Special Publication 1007*, National Institute of Standards and Technology. Gaithersburg, MD, December 2003.
14. Cleary, T., "Characterization of Residential Nuisance Sources," NIST Technical Note in preparation.
15. ISO/FDIS 13751, "Life-threatening components of fire – Guidelines for the estimation of time available for escape using fire data," 2007

www.ingramcontent.com/pod-product-compliance
Lightning Source LLC
Chambersburg PA
CBHW081831170526
45167CB00007B/2786